Quantum Psychology and Mental Quantum Tools

Krystian Defer

...The biggest mountains are in the lowlands....

Introduction

This publication is unique for me. In it I have changed my certain optics of describing Scientific laws not so much from the point of view of physics (esoPhysics), but rather from the point of view of psychology and that of quantum psychology. Of course, there is also a lot of physics here, after all, everything is physics, and mathematics is its language and description, but mainly there is in this publication a strong emphasis on the use of mental Quantum Tools, as ways of assimilating reality and modeling this reality according to this principle ... if a person messes around like this, something always doesn't suit him, and then this can be changed with mental Quantum Tools. And all this is a manifestation of the transition to a new level of our civilization, the so-called civilization of Quantum Man. We are unlikely to build a Paradise on Earth, but with mental Quantum Tools we will be able to take more care of our own health, happiness and success in life. Which I sincerely wish for each of my readers.

1 Quantum psychology. Kickoff

This current of psychology is a novelty, for never before has this science referred so directly to concepts native to Quantum, Cognitive Science and attempts to apply the laws of Nature, Physics to improve the qualitative comfort of human mental life. Such attempts began only in the late 1980s. The official attempt to explain the adequacy of such an approach in psychology within the framework of Quantum Science is to note that quantum laws act on human mental processes, because they are universal laws of Nature, and this is an indisputable fact. Only the matter of explaining this influence crashes into the rather conservative treatment by the official interpretation of Quantum Physics, which is accepted by physicists in general, and which is extremely atheistic and materialistic and single-level. It recognizes as physical facts only what can be measured, i.e., Physicalist processes. This is known as Physicalism. Either the Copenhagen Interpretation of Quantum Mechanics or any of its numerous clones, or the Everett Interpretation, or the Multiverse Interpretation, is considered official. And even such representatives of the Quantum approach in psychology as Joe Dispenza or Amid Goswami do not know how to go beyond the official mainstream of quantum physics. And just to remind you of the fact that single-level interpretations of Quantumism lead by a simple route to descriptive and logical inconsistencies and numerous paradoxes. Be that as it may, such as the Problem of Measurement in Quantum Science, or even Schrödinger's Cat Paradox, or even the impossibility of explaining the phenomenon of Radiesthesia on the grounds of the official current of Quantum Science. One can, of course, overlook and disregard the evident empirical evidence for the existence of subtle energies in Radiesthesia. Only, how long can one? After all, physics is based on empirics, and the fact that the official mainstream of science cannot explain the phenomenon of Radiesthesia, or other paradoxes of quantum physics, does not prove that they do not exist. After all, radiesthesia is pure empirics. Well, one simply has to humbly accept that all one-level interpretations of Quantum with the flagship Copenhagen Interpretation are false and look for some new, sensible interpretation of Quantum. In my opinion, there is such an interpretation and I have included it in my author's esoPhysics. The legitimate question is: why after more than a century since the discovery of Quantum Mechanics still, at least the official mainstream of science, has not developed a correct and true Interpretation of Quantum? Answer: because the mathematical formalism of Quantum is flawless and physicists, guided by David Mermin's famous call: Shut up! And count!!!, the formalism alone is completely sufficient, and they exploit it unreflectively at every turn, leaving the Interpretation of what they do to philosophers. Unfortunately, the effect of this is lamentable, and unfortunately scientists, not only physicists, do not really know what kind of World they live in. Therefore, every now and then, I will smuggle in this publication the basics of esoPhysics, that, in my opinion, the only sensible interpretation of Quantum. But, rest assured, this will not be a book like my "Cognitive Prothesis" where I explicitely presented more-or-less skillfully all of esoPhysics. It turns out that it was probably a bit too difficult for a moderately knowledgeable reader of Physics. However, in order to explain how Quantum manifests itself in the procedures of quantum psychology, I will have to touch on some topics. First, man, as part of Nature, must use the means that belong to that Nature. And all physical processes are carried out under the influence of Causes, which cause certain Effects. In the aspect of man, this Cause (Causal) is his Free Will. Otherwise man cannot act, influence the World. Immediately thrown in here is the vision of explaining the role of man as an "Observer", a role that is so accentuated even by proponents of quantum psychology or even by Quantum theorists. Well, okay, Observer, observation, but what does this really mean in the physical process? It is some kind of interaction with the quantum system, the very one being studied. But the interaction in physics is manifested by a specific Cause (Causal or Purposeful). And what is the Cause here in the case of observation? This Cause is the Free Will of the "Observer". According to esoPhysics, it is worth mentioning here that the Free Will of man and the Will of the Absolute are two new types of Elementary Forces, or Causes (Causal and Purposive), not considered by Science so far. Normally, here the Hard Matter Free Will of man is realized by ordinary Work. That is, according to Free Will, buildings, machines, etc. are created, but it turns out that man's Free Will can also be realized differently. In order to explain this completely, however, I must refer, at least in part, to the theory of esoPhysics.

According to esoPhysics, the World is divided into two levels. There is that level that surrenders completely to Physicalism, the material one, called the Measurable Level, but there is also a second level of the World, slightly different from the former, that level that does not surrender to measurement and observation. In esoPhysics, this second level is the Unmeasurable Level. It is the level of the Spiritual, of Transcendence. In my books, especially in "Cognitive Prothesis," I prove that on this second level the Will of the Absolute is realized. The Absolute (God) has also given us human beings, rational beings in general, an attribute such that they can realize the manifestation of their Free Will on this second level, on this Unmanifested Level. How does this happen, on what principles does it work? I refer you to my earlier books, especially the book "Cognitive Prosthesis". And it is this- the realization of our Free Will at the Unmeasurable Level- that the ability will serve us in application in the construction of the methods of quantum psychology including mental Quantum Tools. Let us also add that the center of a person's Free Will is not located in his brain or body. This is empirically proven by numerous neurological and physiological studies. Normally, from these studies, it follows that we either have no Free Will and are automatons, or.... Well, just either the Free Will Center is an attribute of our Souls, i.e. emanations on the Measurable Level of Spirit, which belongs to the Non-Measurable Level. Since Free Will is an Elemental Force of Nature it is therefore a primary concept to our bodies and our physicality, so it cannot depend on our human physicality (animalness). However, it is also worth realizing, which unfortunately is not communicated by the official mainstream of Science, Physics, that the whole formalism of Quantum is a description of Physics, or rather, esoPhysics, from the level of the Unmanifest. It is not surprising then that a failure to understand this, this basis, this base of physical analysis of the quantum system, leads to so many errors, paradoxes and logical and descriptive inconsistencies. In order to clarify this issue a bit, without going into more detail, in this book I will limit myself to the fact that the mathematical formalisms of Quantumism (Schrödinger's Image, Heisenberg's Image) are written in the spirit of complex numbers and functions. And as you know, composite numbers and functions are unmeasurable. Doesn't that tell you anything already? Well, it does. What it does tell you is that these formalisms are descriptions of physics (esoPhysics) from the

Unmeasurable Level, and almost all concepts associated with these formalisms belong to this Unmeasurable Level. The trouble is that the official mainstream of Science, Physics completely ignores this. And even more, it ignores and mixes entities from the Non-Measurable Level, such as: state function of a quantum system, superposition of states, quantum entanglement, with concepts native to the Measurable Level, such as: Eigenvalue Measurement. And from here, for example, comes the Measurement Problem, or Schrödinger's Cat Paradox, for example, and many other errors. We, however, knowing this, will try to use it precisely in Quantum Psychology by constructing Quantum Mental Tools. I personally am an amateur physicist myself, so I focus the problems of modern man through the lens of physics, in this case, esoPhysics. I make no secret of the fact that my own problems, including mainly health problems, determined my "discoveries" in the field of quantum physics. Consequently, they must have also led me into the territory of modern psychology, and how quantum psychology. Well, because, after all, our problems, that is, our life sculpts our character, our Soul. So, before I go further in constructing tools to help quantum psychology to improve our lives, our mental and physical well-being, let's step back a bit and define the basic facts about man and the human psyche in general.

It is known that we are born in a physical body after fertilization with the sperm of an egg, i.e. a Zygote is formed. However, we actually become a rational human being only with the moment when our Soul, that emanation on the Measurable Level of our Spirit from the Non-Measurable Level, "flows" into the body, "embraces" the body. It seems that the brain is to some extent the "receiver" of the Soul. How is it done, where is it done? This is not known exactly. For the time being, Science is not yet very interested in explaining this. Why? Because modern Science is extremely materialistic (see: Measurable Level) and atheistic (see: one-level), and it doesn't care, ba!, it even bothers to accentuate the Spiritual character of the World. Esotericists and occultists have and have had suspicions for centuries that the pineal gland plays a major role in this process. But these are only suspicions and it is not even that important whether they are true. What can be said for sure for today? Namely, that human consciousness is a superposition of the human Soul and its animal, physical properties of brain structures, neuronal structures that make up the physical brain. I will try to prove this in a moment. It is known that any real entity, that is, one that manifests itself on the Measurable Level. That is, a mountain, a stone, a house, a hospital, a train station, a stick, in a word, every material inanimate, but also animate object, in addition to its typically Measurable properties, such as: weight(mass), charge, also has prices typically from the Non-Measurable Level and only occurring there, such as: spin (that's an elementary particle), energy from the Non-Measurable Level, that is, energy that cannot be measured by measuring devices. And a stone, and a stick, and a mountain, etc., has a certain energy expressed in Bovis units, which radiesthesists can determine. Or measure it? After all, that would be, in light of what I write, a paradox? And here we come to the clou of proof. Because man himself is partly a part of this Unmeasured Level, if only through his Soul. So, he doesn't need some external measuring apparatus to record this energy on the Measurable Level. A simple pendulum is enough for him to register the subtle vibrations of his muscles as a reaction to this energy, let's call it inner entity energy, or Prana, or Chi, or Ki, or Mana. There are a multitude of names for this type of energy, and it depends mainly on the region of the world in which such a name operates. It should be noted here that Asia has already overtaken Europe in this regard for centuries, millennia, and there, in Asia, it is no longer discussed, but considered an objective thing. Moreover, even modern neurology or cognitive science is already loudly emphasizing that the phenomenon of consciousness cannot be explained in the field of materialism and atheism, in the field of mere physical work of the brain's neural structures. There must be something more here than mere matter. There is a qualitative leap from the level of the meta description of the neuronal workings of the brain to the level of the feeling of our consciousnesses, by me, by you, dear reader. This is what no materialistic and atheistic Science or Ideology can jump over. It turns out that Descartes' division into the Spiritual and the material (that which is animal) pertaining to the existence of man now in the light of esoPhysics is confirmed. And indeed, in the light of esoPhysics and from historical and traditional accounts, if only from the entire legacy of Esotericism, which I propose not to ignore, it appears that parents in the act of sex and love do not create a child, in the sense of a human being, but create a body for that human being. Each Soul (actually, Spirit) of this Non-Material Level cyclically incarnates, and with each act of incarnation follows a certain Path of Spiritual Development, defined for it. For what purpose? There are several opinions, but one of the more certain ones is that a person on the Measurable Level, in these cruel conditions of existence, has a cycle of life lessons (and exams!) to overcome, in order to finally reach such "perfection" as to free himself from this samsara (cycle of incarnation) and reach the level of nirvana, i.e. eternal, under the conditions of the Non-Measurable Level, happiness. Of course, no one really knows God's intentions and how or what that would mean. The reason is that no one has returned from the Unmanifested Level and given people any testimony. But that's what logic would dictate. Logic and Moral Laws, which, however, in spite of- I can already hear it- the protests of some people who are enervated by the life of the fauna and flora, and believe that from this point of view that only the cruel struggle for existence is important, just as among animals, which, however, these Moral Laws are quite clear to any sane person. Indeed, the laws of fauna are cruel, but animals are at a lower level of Moral development. They murder each other and inflict suffering on themselves in order to live, according to the principle: the bigger eats the smaller. However, let's leave the cruel earth fauna. For man is on a higher Moral Level, and man is harmed in the literal sense (to his health), and figuratively (to the plight of life) by violations of Moral Laws done by him, or simply by sinning. This is perfectly reflected in the Generalized Law of Karma, presented, for example, in the theory of Dr. Eng. Jan Pajak. Which boils down to the fact that for every sin a person is responsible already in his lifetime. It was Dr. Pajak who pointed out that we, with our illnesses and quality of life, pay for our sins already in this present life of ours. I have added to this Pajak theory the concept of Ethical Implication, which explains what doing sin consists of and how it lowers our Moral Energy Level, which is the decisive factor of what kind of life we lead. I encourage you to read my publication "esoPhysics" where this is discussed in more detail. It is esoPhysics, not a book just an attempt to describe the Physics of the World, that gives credence to this cycle, called samsara, of every human being. After all, the Spirit of a person from the Unmanifested Level is, one can imagine, some form of Energy, and Energy is eternal. So, if a person fails in one life, he or she will probably have a chance to "fix", "what he or she has broken". Of course, this is just a guess, because Judeo-Christian religions, for example, shun the concept of samsara, believing that the Last Judgment and Divine Punishment await those "who have failed" in their lives. According to the messages and esoPhysics, we, as Spirit, choose for ourselves the body, parents and conditions, at least the initial ones, of life on Earth. Interestingly, some people emphasize and that we choose for ourselves a planet and a race of intelligent beings, where we will incarnate (for example, so claims Robert Bernatowicz). And this is very likely, for certainly life and evolution did not develop only on Earth. The Cosmos area is so vast that the races of God's

Children, such as humans, are multitudinous. And rather certainly evolution proceeds similarly on all planets, for such are the laws of Physics. The differences between the Children of God, or races of intelligent beings, are due to the technical progress achieved by these races and the progress of the Moral level they have already achieved. Thus, if we consider the progress in Moral Development, which must also accompany the progress of the species and races of Intelligent beings, then from this point of view the so-called Star Wars idea presented in the Star Wars film series can be rejected. Neither the Empire of Evil exists, nor will it arise, for with Moral progress there is accompanied by increasing responsibility and punishment for acts committed against the Moral Laws. In humans there is also this progress, but for the time being we stand, as far as the Moral level is concerned, only above animals. This is my personal reflection, but it seems to me that our Elder Brothers, the other intelligent races, are not revealing themselves to us because of this relatively weak Moral Level of ours. And remember, we also have a weak and mediocre level of civilization, our Science has a gross five hundred years of development, of course, Modern Science. On the scale of millions of years of evolution, this is not even a second. So our current concern should be, not fear of aliens, but fear of AI (Artificial Intelligence) getting out of our control, which could have disastrous consequences. Technological progress must be accompanied by Moral progress, as evidenced by our evolution, our species. It's just that for most people, let's admit the primitive ones, this is not yet as apparent as it should be. However, morally sick, primitive people, perhaps blinded by their strength, health and vitality, and animalism are not sufficiently aware of this.

As for quantum psychology, here we need to consider and analyze more the structure of our brains in the context that our consciousness is defined by the superposition of our Transcendence, or Soul, with our animality, which is characteristic of the structure of the brain and our evolution in this Darwinian sense. Yes our Souls, our Transcendencies "include" our bodies, but bodies are a product of evolution. And the brain, its structure and functions are a consequence of Natural Selection.

Certainly from our ancestors we inherit the body, its structure, its parameters. Beauty, musculature, bone structure, to a large extent our height. But above all, we inherit the brain. A brain that is the result of evolution from reptiles, mammals all the way to humans, and that incorporates this evolutionary path in its structure. Yes, certainly, our bodies, partly mental properties, are the result of Darwinian selection. But the interesting thing is, are our diseases also the result of our ancestry? It turns out that only about 1% of diseases are genetic in nature. Most diseases are earned by the stresses we experience, our own habits, eating habits, and neglect of the body's physical condition. It's true that in these unfavorable circumstances of our neglect, most often diseases awaken and develop, to which, so to speak, we have a certain tendency from our ancestors. But it is precisely the typical genetic diseases that are so relatively few, for it is Nature herself who has seen to it that for us the most attractive sexual partners, that is, those with whom we will beget our descendants, appear to be such individuals who are genetically extremely different from our own genetic pool. This is all so as not to duplicate genetic errors in our chromosome. This may seem shocking at first glance, but these are the facts. However, it does not change the fact that our parents are only the "creators" of our bodies and certain character traits (and even significant ones). However, our Souls have a pedigree of Transcendence and, in light of what is established by esoPhysics, are the work of God. Without the Soul, the physical body dies. The Soul is such a Homunculus in our brains. In the case of man, there is a superposition in the brain, or rather, his consciousness is a superposition, or in other words, an overlap, of the qualities and characteristics of the brain resulting from natural selection and evolution with the Soul of man, which is an expression of the manifestation on the Measurable Level of the Spirit of that man from the Non-Measurable Level. These traits from selection are all that is studied, brain neuroscience, cognitive science, behavioral psychology. In our brains all the time, there is a dynamic sculpting of neural structures by the turns of our own lives, our stresses and the good ones (eustress) and the bad ones (distress), our emotions and feelings. To paraphrase a well-known slogan straight from the General Theory of Relativity: our neural structures cause certain of our behavioral reactions, which in turn behaviors sculpt our neural structures, and the loop closes. But is it so completely? Well, no. Well, a deeper meaning can be found in all of this. This process, this sculpting of our brains, our behaviors and characters constitutes the Path of Spiritual Development, which is the proper purpose of our being here on Earth, here on this

Measurable Level. In fact, this Path of Spiritual Development is about sculpting our Souls and our Free Will, which is an attribute of the Soul. And we will be held accountable for this whole process after the death of the physical body. We are afraid of death, everyone is afraid of death. That moment when we leave our corporeal shells, our bodies, and what will happen then? As I wrote in my earlier books, especially in "Cognitive Prosthesis," where I presented the physical evidence of the existence of the Transcendent and Immanent God, and the whole concept of the two levels of Reality, it seems that there is nothing to be afraid of that moment of transition. Well, but where? To the Immanent Level. And while I am convinced of the validity of esoPhysics, I too am afraid of death and what happens "later". That's why, and this is the main reason, I discovered and developed mental Quantum Tools, and that's why I try my best to make this mortal life easier just by using mental Quantum Tools here on the Measurable Level. Free Will, which is an attribute of the Soul, is at the same time an Elemental Force that has not been discussed and considered as such by the official mainstream of Science so far. The Elementary Force of Nature, the Source Causal Cause. On what basis do I make such a claim, what authorizes me to do so? I realized this conclusion after an in-depth logical and physical analysis of how the Two-Point Method works. The one just discovered by Dr. Bartlett. For all this to make sense, for the mathematical formalisms of Quantum, Formal Logic to agree with it, and for it to interact with the Pythagorean Principle, it must be so, because it cannot be otherwise. On top of that, I have come to formulate my own proprietary Mental Quantum Tool, which I call the Philosopher's Stone Algorithm, and based on the fact that its operation is empirical proof of this, I can claim that Free Will is an elementary Causal Force, a Source Causal Cause, which can operate under certain conditions even on and through the Unmeasured Level. And this Free Will is at the disposal of us, specifically our Souls. Since Free

Will is at the disposal of the Soul, i.e. the manifestation of Spirit on the Measurable Level, it is not surprising that the Will center in the brain operates with a certain delay than it seems it should. The cursory conclusion from this fact so far has been, this is how Science interprets it, that Free Will does not exist. And interestingly enough, this is the official, though one must admit rather awkward and widely concealed, position of neuroscience. For the sake of completeness, we can add that the official position of atheists on the interpretation of Quantum is the Everett Interpretation. It is characteristic that atheists dominate mainstream Science. They prefer to use falsehood and nonsense rather than admit the Spiritual nature of the world.

The human brain is a creation of many millions of years of evolution, from apes to humans. All of one's abilities necessary for life at the Measurable Level are the result of evolution. All five (for simplicity of argument, let's limit ourselves to these 5 classical senses) senses are the product of evolution and natural selection. Also memory, psyche, emotions, the general character of man. Even his intelligence. But the basic core of consciousness is the Soul. She belongs to the Transcendence, she survives the death of the body. Belief systems based on reincarnation and the wandering of souls through incarnations emphasize just that. The evidence, of course, is not there, because there is no return from there. There is only a one-way road across this Measurable Level - Non-Measurable Level barrier. Once we lose This body, there is no return to This life, in This body. And no one from there has returned in the same body. But we, knowing the mathematical formalisms of Quantum and knowing the correct form of Causality, can nevertheless establish something. So, referring to esoPhysics, I can say that it confirms, as it were, reincarnation and this tradition of the wandering of souls (samsara), and this laborious from life to life refinement on the Path of Spiritual Development. Because since the existence of

God is proven and this division into two levels of reality is real, it must be so for it all to make sense. Atheists differ from Theists in that they believe there is no God, and the world makes no sense. We prove that there is a God, and the world hence has a meaning to its existence and human life also has a deep meaning. As we have already established, the brain is, this physicality and materiality of it, a product of evolution. As I have also written before, the neural structures of the brain are dynamic and change with every feeling, every emotion, every thought. This sculpts our psyche, and it influences and changes our Soul, the Transcendence within us, to finally determine how we have walked this whole life path. Moreover, our Free Will also undergoes this transformation. In the course of life, a person shapes his Free Will. And in life there is no shortage of pernicious habits, addictions, unfavorable habits, that is, those elements that lower our Free Will. And how we cope with this, such will be our "carved" Will, such will be our carved Soul. Beautiful or

Hideous? Will we liberate ourselves from samsara, attain nirvana, or laboriously continue to be a participant in this life relay? What is peculiar is, memory is probably not an attribute of the Soul, because in reincarnation from body to body the Soul does not take with it the memory of past incarnations. This real memory, because it only takes with it this Karmic Memory, so that we pay with our fate for the sins of past incarnations, but we are not aware of this. Therefore, this is the main reason that people generally do not remember their past incarnations. Only hypnotic regression can lift this mystery, but this is already, this is my opinion, from the Akasha Chronicle. During hypnotic regression, we can have a glimpse of the Akasha Chronicle and get information from there. The Akasha Chronicle is otherwise known as the Divine Energy Matrix, on which are written all the events that have happened and all those that may happen. Contrary to appearances, the Akasha Chronicle confirms that the future is not determined from the Point of View of the

Measurable Level. Because what can happen is not equivalent to what will definitely happen. The future is not determined, but it is not random either. Such a small paradox, but it stems from the fact that it is God, His Will, that Determines from the Unmeasurable Level what will definitely happen in the future. However, it is not predetermined. It happens on an ongoing basis, so to speak. This is because God is Immanent. I know, atheists prefer to adopt the "belief" that with each process there is a multiplication of Worlds, but "God forbid!" there is no God. Really, Dear Reader, does this seem plausible to you? They prefer to assume so, rather than believe that the World is a Conscious Design, and that God, His Will, is an indispensable part of physical processes. Atheists prefer to think this way with the intention that humans are autonomous and not dependent on anyone, on the other hand there is a question that has no satisfactory answer: Who created God? However, science itself has come to the conclusion that there is no answer to certain questions. In order not to be lip service, I will give a question of this nature that has no answer: can you give the full expansion of the number Π (the non-measurable number Pi)? Well, there is no answer to that. And if people are autonomous and independent, then they are also, according to such logic, capable of anything. And to lofty deeds and to the latest meanness.

Before I go straight to quantum psychology from the point of view of esoPhysics, let's deal for a moment more with the resilience of our psyches. One thing is certain, since consciousness is a superposition (overlap) of the Soul with our learned and evolved (animal?) skills of the neural structure (brain hemispheres), it follows that it is possible for another person to at least destroy all these learned and evolved skills. All these "faults" of the brain have been parlayed for many years in neurology, brain physiology, cognitive science or simply psychiatry. That is, it is possible, in other words, to destroy and damage a person's physical brain tissue. Well, after all, it is the body, and the body is just a thing, a material object, that is, impermanent. The findings of science in this regard are impressive. All the centers belonging to our senses and abilities are known and well localized. The Central Nervous System is very well understood. Both the brain itself and all the nervous systems. Their role, their importance, their diseases and dysfunctions. The connection of the Brain and nerves with the entire body, with the internal organs, is well learned. This is somewhat overshadowed by the complete lack of Science's position in relation to the distribution of bodily energies from the Non-Measurable Level (Prana, Mana, Ki, Chi) in the meridians and Nadis of our physical "bodies."

— PRANA —   — MANA —   — CHI —

Science ostentatiously disposes of this issue on the grounds that it conflicts with the atheistic and single-level Interpretation of Quantum that prevails in Science today (the Copenhagen or Everett Interpretation). It is also known that the brain is cumulative in nature. That is, it accumulates everything, consistently, but precisely commutatively. It also means that there is a certain lithium of accumulation of the brain's abilities, followed by painful overkill. This applies to: experienced stress, emotions, feelings and even just knowledge. There is a common belief that the brain has "infinite" capacity. Which, however, is not true. Not only too much stress, emotion or feeling can kill, and that in the literal sense or lead to real mental degeneration. A person can also be destroyed by an overload in this regard, that is, too much accumulation, which is done over time, in life so day by day, so god forbid. Psychologists then speak of distress, which, if it turns into chronic distress, can ruin a person, his psyche. Until recently, there was no "cure" for this, no remedy. If a person crossed a certain line in distress, he was practically finished. But now, in these days of Homo Sapiens Quantum, there are already some methods of quantum psychology, which can be called mental Quantum Tools, that can help. I, for one, do not claim that this is an absolute cure, but one can at least try to help oneself, even in those formerly hopeless cases. I myself am the creator of such a Quantum Tool, which I called, I admit, rather deliciously, the Philosopher's Stone Algorithm. And how does it relate to the Soul, this Transcendence in us, in people? Can it also be destroyed? I answer just as our body has to do with our entire life. If someone is stubborn then he can kill, injure and even destroy someone's psyche, after all, Free Will, his Free Will, is an elemental force. And in this way one can destroy a person, his psyche. But also, his Transcendence? This I do not know, I would bet that everything can be destroyed, but the Soul, the Transcendence probably not. It can be changed, after all, our Path of Spiritual Development is about that. I personally doubt this ultimate, such a Christian, Last Judgment, and that supposedly after the Judgment, when someone "flops" it, he is annihilated, that is, his Soul (specifically, his Spirit from the Immaterial Level) annihilated, so claims Pope Francis. I rather opt for samsara. One can only ennoble or debase the Soul, but not destroy it "physically."

According to the traditional division, the human psyche is divided into the Subconscious and Consciousness. In the past, the division into the Superconscious was still emphasized. Such a division is the spirit of the Huna. But today it is unanimously believed that this Superconscious is part of the Subconscious, that is, processes that do not happen in a conscious way. In the work of the Subconscious, the true computing power of the brain becomes apparent. The processes of the work of the Subconscious control the hidden work of the internal organs and those mental processes that take place in parallel, non-linearly. Under these conditions, the work of the Subconscious can only be compared with the work of a quantum processor.

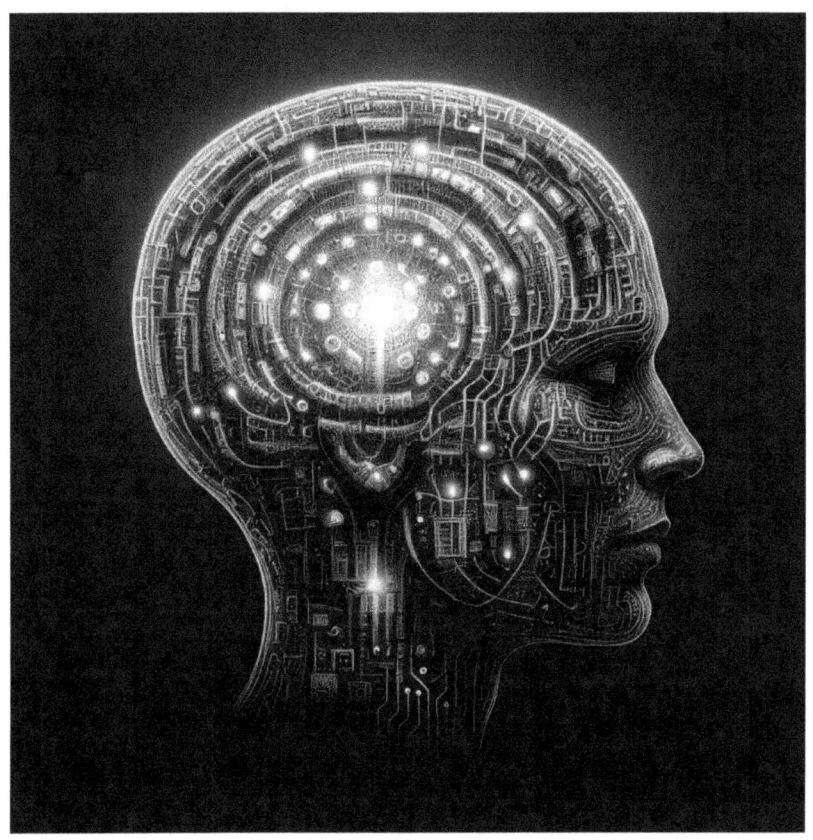

 A quantum processor, on the other hand, is an arrangement of quantum entangled cubits, in this case bio-cubits. How is this realized in the brain? This is something that Science has not yet investigated and defined. The obstacle is to treat Quantum in terms of one-level categories, such as the Copenhagen Interpretation does, which is as of today. I can describe it in terms of the Theistic Interpretation. And according to me, there is a quantum entanglement of parts of the bio-cubits in the brain, which after this entanglement "pass through" and work on the Unmeasured Level, working there in the likeness of an ordinary, technical quantum processor. A working n-cubit quantum processor (n cubits), on the other hand, works at once on 2^n computational processes. If we have 50 qubits, and they are all quantum entangled, the computing power so understood is 2^{50} computing processes at a time. Immediately here comes to mind such a comparison that, as Leibniz emphasizes in his considerations, God at once has the ability to influence all the physical processes that happen in the World. He does it, as it were, at once. And here this reflection comes to mind, that, as the example of the quantum processor proves, similar multitasking is possible at the Non-Manifest Level, and God works mainly at the Non-Manifest Level. In these terms, let's consider what kind of multitasking God, the Absolute, has. Let's calculate. If the particles in the observable world are about $N=2^{80}$, this is only an approximation and an order of magnitude, but it is worth noting under these conditions 2^{79} particles would be half the World!, then the number of simultaneous physical processes must be proportional to all the mathematical relations of this set of particles, that is, from Combinatorics it follows that the multitasking of God would be of the order of 2^N. This is a huge number, but not the largest number known to humans. The text editor doesn't cover this, so I'll write in words. This number of operations at once that God operates is of the order of two to the power of two to the power of eighty. This seems unbelievable, but the example of the quantum processor shows that it is not impossible. All the more so because at the Non-Measurable Level such multitasking is possible and rather ordinary. That is, it can be concluded that our Subconscious is working under conditions of multitasking with enormous power (of course, not as enormous as God) of some (what?) set of bio-quantum processor. This work of this quantum bio-processor, which is probably the Subconscious of man, allows the brain to control almost all physical processes in which it participates, which affect it, every human being. This includes the work of almost all the cells of the body and all the sensory stimuli that man receives. And also, his entire "hidden" psyche. Man's consciousness, on the other hand, works linearly, without multitasking, logically, and relies on what is suggested to it by the Subconscious. A conscious person can effectively be focused on practically only one activity. This is such a characteristic difference between working on the Unmeasurable Level (cubits, subconscious, Absolute) and working on the Measurable Level (Our Consciousness, ordinary work). At the Non-Measurable Level, a locality reigns, everything is interconnected, composite, there, as the mathematical formalism of Quantum shows, the physics there is wave-like, governed by composite wave functions, hence, acting directly there (quantum entanglement), we act on everything at once, according to multitasking. On the Measurable Level, on the other hand, physics is local. Consciousness acts mainly on the Measurable Level; hence it is local and lacks multitasking. The mathematical formalisms of physics at the Measurable Level are "inferior" because it is the Non-Measurable Level that is the Source Level for Source Causes. Therefore, Classical Mechanics is "inferior" to Quantum Mechanics. This is because Quantum Mechanics, Quantum's in general, considers Source Causes from the Non-Measurable Level, which cause specific Effects, already manifested at the Measurable Level. Let us note that our Soul is from the Measurable Level, because it is a manifestation of Spirit from the Non-Measurable Level, but just manifested at the Measurable Level. Hence it is easier for us to explain why our Consciousness, which includes our Soul, is also largely from the Measurable Level. Hence the locality of consciousness and the lack of multitasking. Hence, "normally" from the Consciousness level we have no insight into the Non-Measurable Level. But under specific conditions this can be disrupted, as empirically evidenced. Actually, also the work of our Sixth Sense, that domain of the Subconscious, testifies to this. Through the Sixth Sense, or Intuition, we can have insight into this Unmeasured Level. For example, through Intuition, some people - I would argue that everyone can - can have insight into the Akasha Chronicle and derive from it the knowledge today called Esoteric.

2. the traditional view

Normally, in this decades-long history of quantum psychology, traditional meditation, heart-brain coherence meditation, visualization, affirmation, resonance of visual and auditory stimuli, etc. are considered its basic tools. What is worth noting here is that esoPhysics has a number of modifications to offer, which I will take the liberty to discuss. This will include quantum meditation, quantum coherence meditation of the heart and mind, and among others, quantum resonance of visual and auditory stimuli, etc. In all of these methods, esoPhysics proposes the use of quantum entanglement as that bond of quantumness that brings these methods to the Unmeasured Level. Of course, in traditional terms, they also took place at the Unmeasured Level, but to an intuitive degree. In this they are and were similar to the operation of the Two-Point Method, which is an intuitive application of quantum entanglement, rather than the

Philosopher's Stone Algorithm, which is already a conscious application of quantum entanglement. And just as this Two-Point Method is an Abacus to more modern mental Quantum Tools, so this traditional quantum psychology is also such an Abacus to the methods proposed by esoPhysics and modern quantum psychology.

On what is the advantage of these methods of esoPhysics over traditional methods? On what is the advantage of conscious action from intuitive action. Well, the effect may be similar, because even a blind person can succeed in finding a needle in a haystack. But you will admit, dear reader, that conscious search and action is much more effective. And, importantly, it always, like the Philosopher's Stone Algorithm, works. For example, a needle in a haystack is easier to find with the conscious use of a magnet than by searching blindly. And this is what esoPhysics proposes.

So, let's take a look at how such traditional meditation takes place, considering it, as it were, from the point of view of esoPhysics?

However, let me remind you at the outset that through quantum entanglement we, that is, those who apply these methods, operate, as it were, from the Non-Measurable Level, which is the Source Level for all physical processes whose Effects manifest at the Measurable Level. By the same token, we can build and model Causes at this Non-Measurable Level, which will cause beneficial, most often health-related, consequences already in the observed reality, i.e. at the Measurable Level. And, it is also worth noting what I wrote about, there at the Non-Measurable Level is multitasking, so we can model many things at once. And on the Measurable Level, for example, when we build a house, we have to put up laboriously brick by brick the whole building (it's about one bricklayer and not a whole team). That is, then we cannot use multitasking. This is already proven empirically, because this is the principle on which the Philosopher's Stone Algorithm, discovered and developed by me, works. Let us remember that God, the Absolute, has given us practically only one method for conscious modeling of Source Causes at the Unmeasured Level. That method is the proper use of quantum entanglement. Yes, we normally also act on the Unmeasured Level, because the two levels are interrelated, but conscious, and I emphasize conscious, that is, planned, modeling is only possible with quantum entanglement. This is probably my discovery, and I was guided to this fact (and to the concept of esoPhysics in general) by Dr. Bartlett's experience, i.e. his Two-Point Method.

So, how do we traditionally meditate? We lie down in silence on our backs or assume some kind of meditation posture and, concentrating on our breath, we observe(mindfulness) our breath and thoughts. We continue like this for many minutes. Seemingly simplicity, but nota bene it doesn't suit everyone, it doesn't work for everyone. What do we achieve this way?

Calming of the Central Nervous System, regulation of all the most important physiological processes of the body and psyche. In a word: sheer goodness. But this does not suit everyone, because it is so passive and just so "intuitive." EsoPhysics proposes something more active.

But let's consider, what quantum process is going on here? The answer, for deeper reflection is obvious, there is an entanglement here of the quantum function of the state of the Heart Field with the quantum function of the state of the mind (brain and nervous system). In this way, there is a synchronization between these fields, which is, as it turns out, of capital importance for improving the functioning of our entire body and our psyche. Perhaps there are other elements of this process that I am overlooking, then, go ahead, I will be happy to learn about these processes for my benefit.

This entanglement is intuitive in nature, but nevertheless real, because the subconscious that directs it is also of the nature of a quantum bio-computer. If a quantum bio-computer, it uses quantum entanglement, because that's what a quantum computer is all about. You will ask: surely, after all, people have been meditating for thousands of years? And I will ask: did Gravity work thousands of years ago? If it worked, then Quantum also worked. This applies not only to meditation, but also to magic, which has been with people for thousands of years. She too, probably relied on the intuitive application of the laws of Quantum.

If we already know roughly what meditation (the traditional one) is all about, let's try to improve it as much as possible. And this is called Modern Heart-Mind Quantum Coherence Meditation. All we have to do is to consciously use the power of the Elemental Force, which is our Free Will, and make the quantum entanglement of the Heart Field with the Mind Field with the coherence of this quantum entanglement, so that the whole process "does not escape" us from the Unmeasured Level.

I know, I know, some will immediately protest, because these are not all types of meditation. After all, there are meditations with the repetition of the syllable "Om" or another word given to us by the guru, there are other types of meditations and there are hundreds of them. But, when you analyze all these types of meditations in this way, you come to similar conclusions as I did. Therefore, I ask you, dear reader, analyze, as a homework assignment, other types of meditations in terms of the quantum entanglement of the components of these meditations. I assure you, you will come to interesting discoveries that will certainly enrich your Quantum Tools workshop.

Meditation with repetition of miracle syllables (OM) is similar to affirmation, and I will try to discuss it further when I write about affirmation as one of those effective quantum tools that modern quantum psychology uses willingly.

These different types of meditation in the course of the development of our civilizations have arisen a multitude, and it is impossible to stop them all. But just this example, which I will present here, can serve as a model and pattern.

So lie on your back or adopt a meditative posture. Calm chillout or wellness type music, or solfeggio frequencies, can flow in the background. Ideally, we should not be disturbed by anyone for those few tens of minutes, but, here's an interesting fact, theoretically with a well-executed meditation of this type it always comes out, it always works, just as the Philosopher's Stone Algorithm always works. Why will it always work? Because it is a conscious use of quantum laws and the Elementary Force of Nature, which is the Free Will of man. That is to say, it is a conscious use of the laws of physics, specifically esoPhysics. When you turn on the light in a lamp in a room, you always expect the same effect. And here similarly.

Modern quantum meditation of the coherence of the Heart Field with the Mind Field.

Start breathing consciously, apply mindfulness. Focus on your main chakras for a while.

[Incant in thought or in a whisper].

By the power of my Free Will, I am now performing the quantum entanglement of the Field of My Heart with the Field of My Mind.

...

{You are actually entangling the quantum state function, the quantum function of the Heart Field with the state function of the Mind. These functions are wave functions from the Unmeasured Level. That doesn't mean you're dealing with a wave, it's just that physics from the point of the Unmeasurable Level is wave. You can see here that the corpuscular-wave duality is not that once we have a particle (for example, an electron) and once we have a wave, it is just that once its (this electron) physics is classical (from the Measurable Level) and the other time its physics is wave (from the Non-Measurable Level).

...

[Incantate on].

By the power of my Free Will, I place this quantum entanglement in the space of the Field of My Heart.

...

{that's where you can maintain the coherence of this quantum entanglement in the maximum way, which is when you operate through the Unmeasurable Level [Dr. Kinslow's discovery].

...

Observe yourself (mindfulness). How your body and your psyche will behave. You will observe a marked improvement in every aspect of your body and psyche functioning. You will become calmer, more relaxed.

...

{When first applying this type of meditation with active quantum entanglement and placing this process in the heart field, some sensations may occur. You may sense a distinct activity of the heart chakra, the heart area, maybe a slight pain, maybe tightness. This is a rather normal reaction. If you have any fears and get overly anxious, then stop, do not use such meditations, but I can assure you that with such a strategy, then you will not achieve anything and soon you will be afraid to even leave the house. In general, life is "harmful" and therefore you need to take responsibility and some risks with life into your own hands, do not blame anyone for your mistakes. But I can assure you, this discomfort around the Heart Field after a few séances itself "passes"}.

...

When you consider that the meditation has been completed. After a few tens of minutes.

[Incant].

By the power of My Free Will, I am now decohering this quantum entanglement of the Heart Field with the Mind Field.

...

{It is rather advantageous to make the completion of the formal screening}.

This is kind of my composition of the Heart-Mind Coherence Meditation, but I encourage each conscious reader to make his own suggestions. If you actively accede to the suggestions for the "exercises" I serve here, and compose them for yourself, you will gain much more benefit from reading this book. You'll gain self-confidence, causal power and the conviction that on your own, without help, you can deal with problems, your own problems. Certainly, you will not shun mistakes and errors, but that's what it's all about. To paraphrase Churchill's words ... do not give up, from defeat to defeat, final to victory ....

In quantum psychology, it is very common to use the so-called Resonance of either auditory or visual stimuli, or the combination of the two types into one, with our mind, with our psyche. It is recommended to listen to the right type of music, with the right harmonics at a certain frequency of vibration. It is also recommended to expose the eyes to certain colors or sights, images. How can this effect on the psyche and emotions of a person be explained? Well, as I wrote physics at the Unmanifest Level is wave-like. This does not mean that entities have a wave nature, only that the behavior of these entities is governed by wave functions. And that sound and light are these specific wave structures, and at a deeper level the subconscious, which registers what we hear and what we see, has the nature of a quantum bio-computer and operates mainly at the Unmeasured Level, hence sound and image resonate, entangle quantum with the workings of the subconscious, and thus influence, resonate and affect and through the Unmeasured Level realistically health-promoting or detrimental with our body and our psyche. But in this case, yes listening to healing music or looking at health-promoting images, it is an intuitive, unconscious action, such a godly one. But! But after all, we, knowing these real mechanisms and having mental Quantum Tools at our disposal, can consciously apply these tools in this case, with which we will strongly increase the effectiveness of such resonances.

Particularly on youtube.com, but also on Spotify and other music platforms such healing music and visual files are offered. Such a file should be opened, preferably accompanied by adequate visuals. And here youtube.com has an unbeatable advantage.

Healing Musical and Visual Resonance with the Heart and Mind Field:

Sit in a comfortable position with an upright spine, use headphones. Turn on a file with the desired music content. There are multitudes of channels with similar files on Youtube.com. Just select the file you are interested in. Let's say you choose the file: Healing the Autonomic System. This is, of course, only an example, it could just as well be file: protection from negative energy or yet another. Apply mindfulness. Feel the music with your whole self, observe the visual content of the file.

[Incant in thought or in a whisper].

I make quantum entanglements of the auditory stimuli coming to my ears and the visual stimuli with the Field of My Heart and the Field of My Mind. Thus, I make a Resonance of these multimedia contents with the Heart Field and the Mind Field.

...

{There are certain Hz frequencies of sound and visual harmonics that act in a certain way on the human body. They can act favorably or negatively. In this type of files use consciously and expose those harmonics that have a specific effect on the human being. As I explained it is, correction: it was, an intuitive action, but just from the Unmeasured Level. A popular saying of Tesla ... everything is frequency, energy is frequency.... And although Tesla was mainly concerned with electromagnetism, but this acumen mainly applies to Quantum and wave physics. Why? Because light is a universal energy, it applies to any type of interaction, and it is associated with a wave, or frequency, its physics is wave. Sound also has wave physics. In fact, the sense of hearing has a wave nature. So the resonance of such stimuli has a very strong effect, with a specific intention. For example, here are certain phonetic harmonics of a certain Hz frequency and their effect on humans:

1. **396 Hz (Ut)**: Helping to relieve anxiety and emotional blockages.

2. **417 Hz (Re)**: Facilitates change and transformation, reducing negative thoughts.
3. **528 Hz (Mi)**: Called "pure miracle," it aids cell regeneration and harmonizes the body.
4. **639 Hz (Fa)**: Improves interpersonal relations and communication.
5. **741 Hz (Sol)**: Helps clear the mind and emotions.
6. **852 Hz (La)**: Supports intuition and spiritual awareness.

These frequencies are multitudinous and do not apply only to solfeggio, for example:

852 Hz -cleanses the subconscious mind

404 Hz - activation of great abundance

417 Hz - removes negative energy from the Aura

432 Hz - heals a person's Aura....

Actually, new harmonics are being discovered all the time and their effect on humans is being determined. Their effect is strong, sometimes a dozen minutes of listening to such compositionally encased sounds is enough to feel their real effect on a person. The more conscious quantum entanglement of these harmonics with the Heart Field and the Mind Field intensifies their effect. It should also be remembered that there are extremely harmful and unfavorable frequencies that harm mental and physical health instead of "healing"}.

...

Chill this visual and phonetic message and observe yourself (mindfulness). Your reactions, your body. You may feel, especially with the first séances of this type, some sensations in the area of the heart chakra. You may also, after several minutes of listening to one type of file, listen to another. However, after the entire session of listening to this type of file, already at the end, make a decoherence of this resonance.

[Incant].

By the power of my Free Will, I decoherence this quantum entanglement (this resonance) of sound and visual files with the Field of My Heart and the Field of My Mind.

In the case of quantum entanglement of two state functions, one with a state (+) and the other with a state (-), and they are continuous functions and not discrete functions like, for example, the values of spins (of electrons), then after some sufficiently long time, these states change so terminally that the function that initially had a state (+) changes to (½+), and the one that had a state (-) changes to (½ -). These are all consequences of spontaneous quantum jumps and the nonlocality of these functions. I will try to explain this in more detail when I write about the Philosopher's Stone Algorithm and the Cognitive Prothesis. Now this is important enough to note that in the case of the Resonance of the sound signals and the vision signals and the Heart Field and the Mind Field, this signal (sound) and the vision signal, to which we can assign a conventional state (+), do not diminish, that is, in the end it is not equal to (½+), because they are constantly replenished on an ongoing basis, because the sound signal does not diminish and constantly has a conventional value (+) and the vision signal also does not diminish and constantly has a conventional value (+). This is what is characteristic of Resonance. Thus, this signal of the nature of (-), that is, the state of the Heart Field and the Mind Field drops by much more than to (½-), as much as to (-1/n), that is, it improves significantly. That is to say, from this point of view, it makes sense to use such Resonance Signals of appropriate sound harmonics and appropriate visual signals. Thus, conscious quantum entanglement of these signals, and not just such intuitive mere listening to these signals, has capital healing properties and enhances this "healing" process.

Let us now make such a **Resonance** in a slightly different way. For example, let's make a quantum entanglement of some **Problem** with a musical and vision signal of a certain healing harmonic, and let's place the quantum entanglement thus made, this Resonance, in the Heart Field. And this is where we have to deal with the fact that after some time the Problem will decrease up to (-1/n) of its initial value, so to speak, we will make a healing. This "n" in this fraction is not easy to determine. It depends on the length of the Resonance session in question, but also on the value of the healing harmonic and vision content itself. It depends on many factors, but ultimately the longer we keep the headphones on our ears, the greater this "n" will be, thus it will cause a lowering (strictly mathematically it is an enlargement of the whole fraction, because it is a fraction with a negative value) of the whole fraction (-1/n), that is, it will also be a better final effect. I will still write about this mechanism itself and try to explain it when I write about mental Quantum Tools.

Resonance of the problem: I lack cash, I am in a poor financial situation with the solution to this problem, that is, music attracts material and financial prosperity.

We open a music file that provides just such a cash injection. There are multitudes of such files on YouTube.com on Spotify.com. Of course, there is no guarantee that they are all equally effective. But once we've made our choice, then we sit down for those several minutes in a comfortable posture, turn on the music file (you can also add video here) and Incant.

[Incant].

By the power of my Free Will, I am making a quantum entanglement of my life problem, lack of cash and notorious complaining about lack of liquidity with the sound stimuli that come through the headphones to my ears.

[Incant].

Thus accomplished quantum entanglement, this Resonance, by the power of my Free Will I place in the space of the Field of My Heart.

...

{depending on the recommended length of listening to this file, which is usually given in the file description, listen to this music for this long. Then finish.

[Incant].

By the power of my Free Will, I am now decohering this quantum entanglement, this Resonance.

{This concludes a single session of this type of Resonance}.

Of course, there is nothing to cite Zeland and his passive action. It is also worth taking some active steps in this direction to enrich ourselves.

So, as you can see right away, a similar Resonance effect will be had by meditating with the repetition of specific sentences, words or syllables, or affirmations. With that said, now the sound signal, constantly amplified, should be replaced with these repeated affirmations, and they should simply be quantum entangled with the Heart Field and the Mind Field, or better yet, with the Problem in question. This will certainly intensify the health sense of such affirmations.

In the case of Resonance, it is worth knowing that there are special channels on YouTube.com, where the files contain both a visual signal, which includes Runes, the Elven Alphabet, Sigils, and elements of Sacred Geometry, which also include a sound signal with corresponding harmonics. And these two signals form a unity. Such files are particularly powerful. One can distinguish such channels as Ankhi Priest or Ankhi Force, or Ankhi Power. These files in these channels are not long, but by virtue of this combination of vision and sound, they are a tremendous power, they are up to 12 min. or a maximum of 15 min. I do not recommend, in the context of the quantum entanglement of these files with a particular problem, to listen to more than four such files in one session. As I reiterate, they are extremely powerful and potent, and an overdose of them can even be dangerous.

One thing to keep in mind, with "normal" meditation or listening to "healing" music, the health-promoting process is done intuitively with a lot of input from our subconscious. With conscious meditation, when we make the appropriate quantum entanglements, then we benefit in the Elemental Force, which is our Free Will. We then act consciously on the Unmeasured Level. You will ask: does it make a difference, is there a big difference? In my opinion, a fundamental one. Such a difference as the difference between an animal and a human.

As humans we can consciously model reality, as animals we are only passive consumers of that reality.

3. stress

In modern times, one thing we have an excess of regardless of age, health and wealth is stress and the negative emotions associated with it. And it should be noted here that this is a rather primitive (early) feature of the brain, precisely the reaction to stressful situations. Its pedigree goes back tens of millions of years, when mammals, or rather their brains, were just forming, but it has proven to be a great evolutionary success. And even the fact that we, in these evolutionary and Darwinian terms, have prevailed over all the rest of the fauna is largely due to this "primitive" feature of our brain, the unconditioned response to stress and stressful situations. But today, today, when we are unlikely to be threatened by a confrontation with a lion or cheetah, when our stressors are already more subtle than a simple struggle for food, dominance or begetting offspring, these primitive brain responses to stressors are the cause of many emotional and mental disorders of modern man. And here is the immediately sad news. Unfortunately, nothing can be done about it, our brain is already rather formed, the structure of the brain can not be changed, that is, in terms of changes over the next few tens of thousands of years.

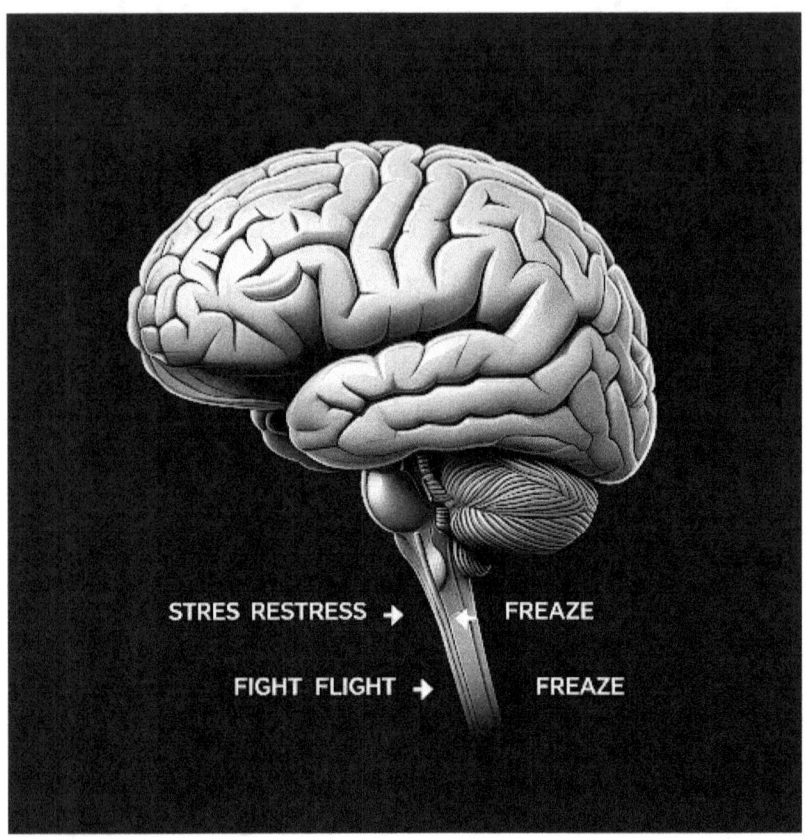

Well! Well, maybe biogenetics will be able to significantly accelerate this process, but still, on the scale of one lifetime, nothing can be done about it. That is, during our lifetime and this being in this incarnation at the Measurable Level, nothing will come of it. But there are at least three pieces of good news that can already give us optimism. First, it's not that the human brain has no adaptation capabilities to stress, that awful distress (negative stress). It even has great abilities to neutralize this worst stress, that is, chronic stress that lasts even half a lifetime. Well, the brain has such defensive skills, and for many people this is enough, but not for everyone. Secondly, over the past few thousand years, that is, when our species has gone through a rather notable path of evolution, no longer biological, but social, and from a form of Stone Age culture has moved over these thousands of years into the era of technical civilization to the present day, man has developed certain tools for coping with stress. I am referring to widely understood medicine, or even meditation techniques, which I have already written about in this publication. All this has allowed us humans to at least partially control this primitive reaction to stresses and stressors. But, of course, this is not enough for everyone. But there is still a third. Third, the era of Homo Sapiens Quantum, Quantum Man, has now arrived. Mental Quantum Tools have now been developed. This is a more fundamental transition than the turn of the 18th-19th century, when the electromagnetic age was born and electricity was liberated. Here, with the discovery of mental Quantum Tools and their use, we have the ability to influence directly the Source Causes from the Non-Measurable Level underlying the Effects from our Measurable Level, so we have the ability to influence the Source Causes underlying our stress behaviors, so we can model our primary behaviors in this context of stress, our stress response, in this context of the stress response of our bodies, and even the stress response of our beings in general. This I will try to describe in detail in the final part of this publication. For now, I'll just write that mental Quantum Tools are a practical realization of the esoPhysics I describe and present in my publications. But, before I get to that, let's go back to the sources and consider, what is this pattern of the brain's response to a stressor and stressful situation? How did evolution solve it? In a stressful situation, any stressful situation, under the influence of a stressor, it is the part of our brains, called the mammalian brain, the limbic system and specifically the amygdala and the hypothalamus, which are the parts of the brain that mobilize the body into one of three responses first. To fight, flee or freeze into immobility. This is accompanied by an outpouring of emotions, usually negative ones like fear, anxiety, irritation, agitation, rage, anger. And it is only after this unconditional reaction, however, that the part of the brain, already more human, that is the prefrontal cortex, as the center of coordination of our consciousnesses, comes into play, that is, only then do we become aware of the situation. But before the prefrontal cortex comes into play, which allows us to assess realistically the stressful situation, a lot happens in these categories of human survival or not. This reaction of the amygdala and hypothalamus is crucial. Someone will say: but you write about Transcendence, about the Soul, and here, as if contradicting yourself. Rest assured, there is no contradiction here. After all, in my publications I write that the body and brain is material, it is part of the Measurable Level. Our bodies and brains have evolved over the course of many millions of years of Darwinian evolution, but even so, all the time there is and has been this bond with the Non-Measurable Level. How to describe it competently and in detail? Well, I probably won't be able to describe it comprehensively, but I have already presented a sufficient body of empirical evidence for just this, and as the science people say: empirical evidence is key.

This response of the body to stress we might not be so important if it were not for the fact that sometimes temporary stress, the kind that saved the lives of our ancestors on the savannah, turns into devastating chronic stress. Chronic stress is already very dangerous, for in extreme cases, years of chronic stress leads a person, his body, his psyche to extreme degeneration and death in suffering. Nowadays, practically stress, chronic distress is the basic form of stress that we humans have to deal with. As I wrote the brain we won't change, neither will its mechanisms. But the key here may be this reflection of psychologists, who say that how we perceive stress in a given situation, that is, whether as distress (negative) or as eustress (positive) in this chronic context, is mainly responsible for **our interpretation of this stress**. That is, however, we return to the prefrontal cortex. But the truth is that first we have to somehow survive this first throw, that is, this involuntary reaction of the amygdala body. And then we already have methods, quite effective: notice, that is, the mental Quantum Tools developed by me, among others. This chronic stress is ruinous. When we think badly about what is stressing us, that is, mainly about what has happened to us, in other words, we misinterpret (prefrontal cortex) the whole situation, that is, we worry constantly, we use ruminations, that is, we constantly roll out the given stress in our thoughts, that is, the problem, we can lead our own psyche and body to lamentable results. And there are times when, in our time, distress can cling to us for years, even decades. So, indeed, it is precisely the case that today we associate stress little with the struggle for survival in the wilderness, but with experiencing life's problems. And from a physiological point of view, these situations - fighting wild animals and experiencing an argument with the boss or a polemic with the mother-in-law - are equivalent to each other. Interpretation! Interpretation! This is the key word, and it proves that actually optimists live better, because they always know how to explain a stressful situation to themselves in such a way as to harm themselves as little as possible. From this point of view, life is better for those for whom the glass is always half full than for those who interpret the world in terms of a glass half empty. It is indeed the case that stress begins through the reaction of the amygdala body, but then it is the prefrontal cortex that takes over. Come to think of it, though, can we really always convince ourselves (the prefrontal cortex) that this stress that's just grabbing us is trivial? Well, not always. Let's admit that even a born optimist, such events happen to him that even he cannot trivialize them. Death of a family member, serious illness, loss of job, loss of livelihood, calamity, traffic accident, etc. It's just that there's still a long way to go from such God-given tributes to real chronic stress and its devastating effects. As I wrote, our brains have internal defense mechanisms that are sufficient for most people to cope with God's tributes in the long term. In addition, medicine and psychology in this traditional understanding already have very effective tools to cope even with high chronic stress, the clinical kind. Well, and then there is this third way, which is quantum psychology, or mental Quantum Tools. This path takes humanity to a new stage in the development of human civilization, to the stage of Homo Sapiens Quantum.

And when it comes to stress, chronic stress, nowadays a fairly common reaction of the body and psyche to this stress is various psychosomatic diseases. In the last century, even coined in medicine such a symptom of the Chicago Seven, that is, a list of seven diseases of the body that are source psychosomatic diseases. These were: Stomach Ulcer, Hypertension, Bronchial Asthma, Rheumatoid Arthritis, Inflammatory Bowel Disease, Hyperthyroidism, Atopic Dermatitis. Today, diabetes, obesity, asthma, atherosclerosis, psoriasis, etc. can still be added to this list. So today, this number of diseases has grown considerably, and it is estimated that a large percentage in the total number of diseases, that is, 20%- 30%, are psychosomatic in nature.

Psychosomatic diseases are an expression of the fact that in a situation of permanent, chronic stress, the weak points, the weakest points of our bodies first "sit down", which initially need not be accompanied by any pathological changes in these points, as a safety valve for the whole body, so to speak. But then, in the final stage, specific pathological changes in these points can and most often do occur. And so, someone for years can have false pre-infarction signals of a psychosomatic nature. That is, he will have symptoms of a heart attack without any organic changes, but in the end, after years, he will really get such a real heart attack.

In addition to specific changes in the body, stressful situations also lead to specific behavioral changes in a person's behavior, in his psyche, and also and above all in his character traits, I mean negative changes. It is not without reason that I wrote about the adverse effects of long-term distress on the body, because this is also directly related to changes in the human psyche. And this is what psychology in general, and in extreme cases psychiatry, deals with. Quantum psychology encompasses all these pathologies at once and tries to "straighten them out" in a quantum way. And those pathologies of the body, and those involving the psyche, the soul. But what types of practical psychology have emerged so far, already after nearly two hundred years since the inception of modern psychology? So, what types of psychology are these?

Here are some of the basic types of modern psychology:

** **Cognitive psychology**: focuses on the study of thought processes such as perception, attention, memory, thinking and problem solving. Cognitive psychologists seek to understand how people process information and how they can improve their cognitive functions.

* **Behaviorism**: focuses on the study of human and animal behavior. Behaviorists believe that behavior is the result of environmental conditioning and seek to understand how the environment affects behavior.

* **Humanistic psychology**: focuses on the study of human emotions, motivation and personal development. Humanistic psychologists believe that people have an innate potential for goodness and seek to help people reach their full potential.

* **Psychoanalysis**: is a theory and method of treatment developed by Sigmund Freud. Psychoanalysts believe that human behavior is

the result of unconscious mental processes and try to help people understand and work through their unconscious conflicts.

* **Social psychology**: studies how people affect each other and how social groups affect people's behavior. Social psychologists seek to understand how people form their attitudes and behaviors toward other people.

* **Evolutionary Psychology**: studies how human behavior and mental processes have evolved over time. Evolutionary psychologists seek to understand how human behavior is an adaptation to the environment and how it can be explained by evolutionary theory.

* **Neuropsychology**: studies how the brain and nervous system affect behavior and mental processes. Neuropsychologists seek to understand how brain damage can affect behavior and how it can be treated.

These are just some of the basic types of modern psychology. Each of these types has its own theories, methods and areas of interest, but they all seek to understand human behavior and mental processes. It turns out that emotions, emotional states and thought processes are key in all these types. That is, as I wrote, in stress, the key is the **Interpretation of the problem (stress, stressor)** and everything related to it. That is, the state and work of our prefrontal cortex is also key. This is perhaps imprecise and does not convey the entire depth of the issue, but to a first approximation this is what it all boils down to. In a stressful situation, in general, in a problem situation, we first have an unconditioned reflex, a release of emotions and feelings, and then the prefrontal cortex **interprets the situation**. In a state of chronic stress, there is a gradual distortion to the detriment of this simple physiological response. So, then these distorted, sometimes in a caricatured way, emotions and feelings, have a destructive power and demolish the very process of thinking, hence the behavioral and mental problems that are a consequence of such negative chronic stress.

It should also be remembered and taken into account that the brain, and in fact everything related to the work of the brain is cumulative in nature. What does this mean? It means that its resources can be overloaded. After all, the brain is matter, he has material resources. And this cumulative nature is progressive, and the brain fills up in the course of life. It is not true that the brain has infinite resources. Although it is true that its accumulation capacity is impressive. Young people may not be so affected by this, but mostly mature people are. The young do not yet know their limitations. They have the potential and enthusiasm to create the world, but once they know their limitations, both youth and potential are behind them, and that's when the brain may find its limitations. What can be overloaded, what can be over-accumulated? Certainly, negative emotions, feelings like: fear, anxiety, worry, irritation, frustration. But also, positive emotions like: joy, euphoria, excitement, etc. It is also possible to over-accumulate knowledge over the course of a lifetime of decades. And in this general sense of the word, and in this sense of specialized knowledge. This accumulation will therefore constitute an overload for our body, for our being. So, it will become a very dangerous stressor, leading in a certain perspective even to perdition and death. And the point of all this is that such accumulations are, without conscious action on them by mental Quantum Tools, practically irreversible. You will ask: and what does our Transcendence have to say about this? Good question. As I wrote, our consciousnesses are a composite, an overlay of our evolutionary brain skills with our Soul, that Transcendence within us that comes directly from the Unmanifest Level. And just as a person can be killed physically, he can be destroyed or he can destroy his brain and psyche, this physicality within himself, which is consequently equivalent to physical death. Can the Transcendence within us be destroyed? I don't know. I suppose not, because our Spirit from the Immaterial Level is immortal, unless the Absolute decides otherwise. What I do know, however, is that it is possible to sculpt and shape our Soul, our Spirit, because that is, after all, the content of our Path of Spiritual Development. And as tradition and transmissions treat it, death under tragic circumstances, the so-called death of a hero, elevates our Spirit. Examples of which are death on the battlefield, while fighting in defense of the oppressed, the Homeland, in self-defense. But no one really knows fully, because there is a lack of feedback from people who have experienced something like this. Because the way there, to the Immeasurable Level in a given incarnation, is only one-way. One thing is certain, anyone who causes the rape of another human being will be held accountable, because that is the Moral Law. In this world or in the Other World. Yes, it is possible to dodge, to wrench, to maneuver, to circumvent even quite successfully the state laws. Statutory laws are raptly the last few thousand years in the history of our species. There were no state laws, there were no codes, no paragraphs and anyway Moral Laws have always protected man. And if someone has an atheistic worldview and thinks that in this way, he will escape justice, then in light of what I write, he may feel some discomfort. Unfortunately, nowadays people fall into a certain moral trap and think that if they somehow manage to avoid criminal responsibility, nothing has happened, and they are as pure as a white lily. Thus, transgression is associated unequivocally with criminal responsibility. However, this is not quite so, because state laws, codes can be circumvented, especially the people who control the state, our lives have such opportunities, but the Moral Laws have not yet managed to circumvent anyone. And so it will come sooner or later to pay for every misdeed that one has committed in life. I suggest everyone to get acquainted with the theory written on this subject by Dr. Eng. Jan Pajak on Human Morality, which I call the

Generalized Law of Karma. Very informative text and explains a lot.

If we already have such a case of head overload problems. Whether it will be related to emotions, life experiences, or even knowledge, it seems that this last resort for a person will be correctly "set" mental Quantum Tools. They are the ones that can change the conditions on the Unmeasurable Level ensuring the return to normalcy of our brains, minds or psyches. As I wrote in the "setting" of such tools, be it the Philosopher's Stone Algorithm, the Cognitive Protector, the whole complex of problems can be attacked, according to the fact that multitasking is the norm at the Non-Measurable Level. Thus, one can "heal" the entire complex of issues at once with this method. I, in my applications of these tools, can spend beaten tens of minutes just "setting up" such a tool as Cognitive Prosthesis. So, I say this, as it were, from the position that empiricism, at least with me, confirms the theory. In general, I developed the whole theory of esoPhysics on the basis of

Dr. Bartlett's experience, his Two-Point Method. And here, too, empirics agrees with theory.

In the next chapter, I will describe how the phenomenon of the inner energy of entities, material bodies, living beings from the Unmanifested Level is manifested and realized through so-called Radiesthesia. And what practical benefits this can bring in the context of human mental and physical health.

4 Strange Energy

Well, that's right, strange energy. Every entity "has" it. And a pebble, and a crystal, and a stick, and a mountain, and a house, and a creature, and even more so a human being. But Science, its official mainstream, rejects these revelations, and although it is strictly empirical, repeatable and verifiable knowledge, Science rejects it. Why? Because the official Current of Science is single-level, and this includes the statement that Science recognizes only entities and bodies, we shall say, from the Measurable Level, the one on which Physicalism dominates. Science recognizes only those entities that are "measurable," that is, those that measuring devices indicate. And this "strange energy" cannot be measured with one exception. This energy can only be measured using the sixth sense, Intuition. So, what is it like? Is it a stitch, or are these things made up? All this Prana, Ki, Chi, Mana, etc. Could the powerful civilizations of the East have been wrong? Could modern Dowsing be wrong and are just telling nonsense?

But all those who recognize that reality is deeper than just what is material must admit that esoPhysics with its division of reality into two levels resolves this apparent paradox.

In the interpretation of esoPhysics, this strange energy is simply the energy that every entity has, even the material one, on the Unmeasurable Level or from the point of the Unmeasurable Level. And even the fact that this energy is "measurable" only by the movement of a pendulum in the hands of esotericisms, and is calibrated in Bovis units, proves that it is indeed immeasurable. For, if it were measurable, it would be measured in Joules or Calories, or other units adequate for energy. Scientists, physicists spread their hands and say that they are helpless, but it is known quite widely even physics celebrities, serious scientists, who secretly, after nights, so to speak, as a hobby, parade radiesthesia or unfold the Tarot.

That is, how to understand that this energy is not measurable, and at the same time the radiesthesia knows how to determine it, ba!, even knows how to determine the degree of its intensity precisely in Bovis units? This ability, that is, to study and locate this energy, almost everyone has, but? But depending on their talent, some do it better, more accurately, others worse, but? But only a small percentage of people are completely devoid of this talent. With only 5% of the population having this defect and not knowing how to do it because of the inverse polarity of their own energy (yes, yes, that internal energy, the one in question). This is because men have this polarity opposite to women. It happens that due to life situations, stresses experienced, we suddenly redirect this polarity. And then we can, and rather should, if we notice this, go to a dowsers who knows his profession properly, so that he can restore this correct polarity for us. The entire operation of this type should not take more than five minutes. Without this we are exposed to a number of unpleasantness. One, that without it we cannot use pendulums and esotericism in general, two, that we are exposed to a number of diseases of various causes, for example, energetic. But, of course, unbeknownst to us, we can live like this with reversed polarity for many years, just as one can live for many years with chronic diseases.

That is, the conclusion is that we humans can measure this unmeasurable energy. Why? Because our Souls source from this Unmeasurable Level. Generally, also because living beings are connected to this

Unmeasurable Level. The phenomenon of life has not been figured out by science to this day, although its efforts in this regard are several hundred years old. And, it can be presumed, Science will never succeed, because life, living beings require that extra element of the Divine, that Miracle of Life, which dead matter does not possess. All this causes that, in addition to the five ordinary (conventionally, because there are many more) senses needed for him to function here on the Measurable Level, man has one more sense, a sixth sense (Intuition), which allows him to communicate with the Non-Measurable Level, with the Akasha Chronicle or Divine Energy Matrix.

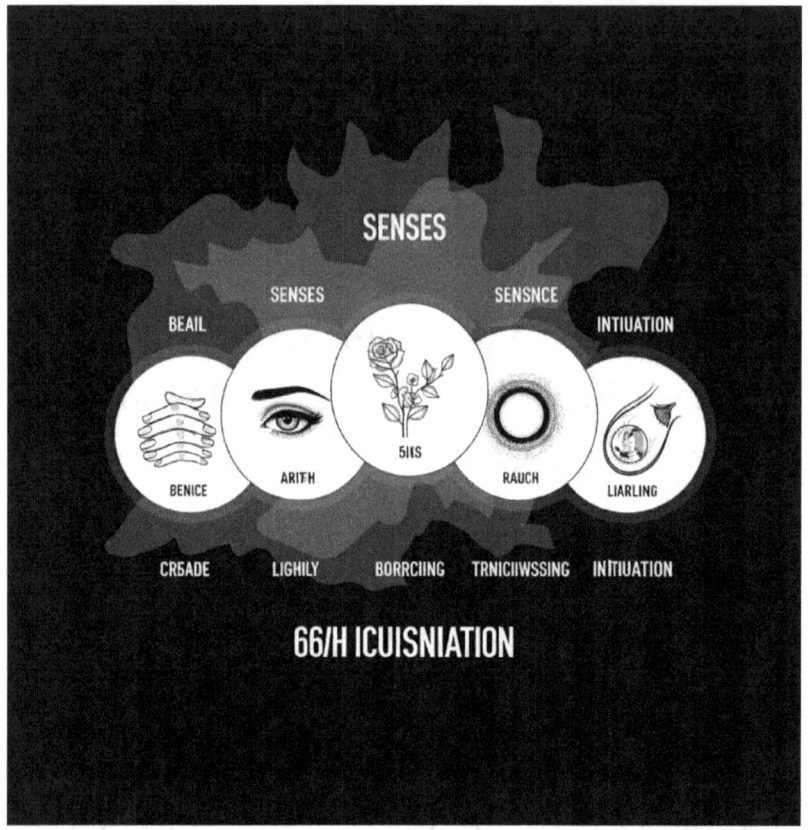

How does it work?

We can perceive this "Strange Energy" through the movement of the pendulum in our hand. The subconscious as this quantum bio-computer works on and receives from subtle information from the Unmanifest Level, including information from the Akasha Chronicle (Divine Energy Matrix), but also from the Unmanifest Level in general, and this manifests itself in the unconscious movements of our hand muscles and also our body, which can then directly affect the movement of the pendulum in our hands. The mechanism is much more complicated, but it boils down to the fact that every person has an additional sense, a "sixth sense" with which he or she receives information from the Unmanifested Level. This is influenced by two factors. Once, the subconscious is to some extent a quantum bio-computer. The mechanism of operation of this bio-computer is not yet worked out, but it is worth knowing that even these physicists of the single-level trend suppose that subtle quantum entanglement processes of structures called microtubules take place in the brain, in and through which quantum processes generate the phenomenon of human consciousness. Two, esoPhysics, and myself included, proposes to recognize that consciousness is the superimposition of brain processes arising from the corporeality and animalism of the brain structure with the Human Soul, which has a connection with the

Unmanifest Level. This seems the simpler (simplest) explanation in the context of all esoPhysics, but how it really is difficult to judge. Only when Science breaks through to the ideas contained in esoPhysics and reliable research in this direction begins, only then will we know the whole truth about this subject. The advantage of esoPhysics is that it recognizes that, yes, the subconscious is a quantum bio-computer, but the processes of the subconscious to the Unmanifest Level are largely unconscious (as, by the way, the name suggests: subconscious). However, man himself, consciously, and here probably with a great deal of involvement of the Soul, can deliberately generate quantum processes (quantum entanglement) with his Free Will, that attribute of our Souls, so as to deliberately act on and through the Unmanifest Level. And both of these processes affect subliminal movements of the muscles of the hands or other parts of the body, which can then be interpreted and judged as the movement of a pendulum or wand.

We can thus perceive the internal energies of material and living bodies. We can locate metal ore deposits, watercourses, etc. this way. We can also in this way draw information from the Akasha Chronicle, where all the events that have taken place are recorded, as well as all those that may, and I emphasize may, take place.

And here such a small digression. If we are dealing in laboratory conditions with an experiment of quantum nature, we can, as a rule, get a set of discrete values of a given observable, and which will be realized, it is already .... , single-level physicists claim, a work of chance (Indeterminism)???? In normal life, if we attach to the known four elementary forces an additional elementary force, namely the Free Will of man, it turns out we will also obtain a discrete set of events that we can get, and which will be realized, it is already .... it already depends on the Will of the Absolute!!! And it is in the Akasha Chronicle that this is recorded. It means that the World is not determined in advance, but it means that God makes choices on an ongoing basis, as it were, He determines the physics of the world on an ongoing basis. This happens on the Unmeasured Level and is subject to the multitasking mode (which I have already described). We will still use this property, this multitasking, of the Unmeasurable Level when constructing mental Quantum Tools such as the Cognitive Prosthesis, to make the most of the power of quantum entanglement. It turns out, and I will continue to write about this later in the book, that this is, these mental Quantum Tools, the most effective protection of a person's mental condition. And the most effective method of solving problems in human life in general.

So, indeed almost everyone has this gift of intuition. Apparently, women are more endowed with this gift than men. It seems that the subconscious plays a key role. This book is not going to be a manual on how to use this gift of intuition effectively, but I will admit the knowledge flowing this way is very valuable and underestimated, especially just by men.

How do esotericisms make the most of intuition? Most often through the work of pendulums. It is the effective movement of the pendulum that carries Intuition's information on a given issue.

As I wrote, this movement is the result of hand and body muscle movements, as a result of the subconscious reaction to a given problem. And there are a multitude of types of pendulums. Each of these types corresponds more or less to a different type of issues related to the Unmanifest Level. Some pendulums heal, others recharge, others purify, others reveal knowledge given only from the Akasha Chronicle, others locate specific entities, etc., etc., etc., etc. However, it is worth remembering that when working with pendulums, it is necessary to apply a kind of health and safety of working with pendulums, otherwise the pendulum goes crazy and gives nonsense. So one should use frequent breaks when working with the pendulum, one should clean the pendulums frequently. However, these are already detailed technical problems of such work. What is known is that reliable knowledge obtained by this means is verifiable, for example, by others using pendulums. So, we are not in danger of such a situation that nonsense and nonsense will be reproduced by esotericisms. Every piece of information is checked and verified. The official Current of Science, although it is strictly empirical knowledge, is helpless here, so Science has employed a strategy here that it rejects everything that is related to Intuition, to the Unmeasured Level, to pendulums. One-level, atheistic Interpretations of Quantum Science, such as Copenhagen and all its clones, are completely helpless here and express their opposition to these issues, as if this would be effective in falsifying reality. The truth will defend itself, as formal logic says, but there is always this possibility of duplicating untruth. And then we are not dealing with logical inference, but with sowing nonsense. So far, it seems that the proponents of one-level Copenhagen-type interpretations have been spreading such nonsense for almost a century.

As I wrote, the skillful use of pendulums is very useful. We will need it when setting up mental Quantum Tools. To, one, properly and correctly formulate the problem, and then, two, properly formulate the solution to the problem. I am writing this, so to speak, in the run-up and, so to speak, in this publication, but anyone who has read my other publications already knows more or less what I mean. So, you can see here at once, what I constantly undertake in my books, that esotericism, magic is nothing more than the effective use of quantum methods. Was Gravity universally applicable in the past? Yes, it was in force and the Earth zapped in its orbit around the Sun. Was Quantum valid in the past. Yes, it was in force and people have this element of Transcendence in them. I'm just showing that magic and esotericism is otherwise an intuitive use of Quantum methods and tools.

5. mental quantum tools

The main basic mental Quantum Tool that I have developed is the Philosopher's Stone Algorithm and the resulting Cognitive Prosthesis, limited to mental and psychological processes, and to the health of the body and soul. Before I go into these tools, however, I must refer to the history of the whole idea, the history of the idea of quantum tools.

Although the theory of Quantum Mechanics is more than a century old, in fact, the real breakthrough was made in this field in the 20th century, in the late 1980s and early 1990s, at which time Quantum Mechanics, broadly understood, became spiritualized. It was then that there was a widespread realization that the prevailing paradigms in mainstream Science, Physics, yes, the atheistic and one-level paradigms, were insufficient and unsatisfactory for a wide range of independently thinking people and people who believed in a deeper sense of reality. Back then, neither thought in similar terms as esoPhysics, nor divided physical reality into two levels. Everything was treated as a single level, and yet Quantum began to be associated with spirituality in the broadest sense. As if subconsciously, in an intuitive way, Spirituality began to be treated together with Quantum, limited at the time only to the idea of the Quantum Field, but it was difficult to reconcile the formalism of Quantum and consider its features as belonging to the same level as Spirituality. It must be admitted that theoretical physicists did not make it easy (they still do not today) to solve this problem. They preferred and prefer today to talk about some hypothetical multiword, rather than referring to Spirituality and the Absolute in their opinions.

Well, in the 1990s, across the Atlantic, a certain Dr. Richard Bartlett, a neuropath by trade, noticed that if one applies a finger of one hand to a diseased spot on a patient's body under certain conditions, and the other finger of the other hand to a particular healthy spot on the same patient's body, a spontaneous reaction sometimes results in healing, and even of severe conditions. Some of his patients were being treated for scoliosis, and then especially often healings could be made this way, and almost completely and the first time. Dr. Bartlett described the entire method in several books and called it, what is still used today, the Two-Point Method. It is also often called the Two-Point. This would have been perhaps just a curiosity, just such a new method of naturalistic medicine, but no, it was actually an expression and manifestation of the realization of quantum laws, of which I guess the creator himself, i.e. Dr. Bartlett, was not aware. For here, along with the Two-Point Method, the first mental Quantum Tool appeared in the world. How to describe this experience of Bartlett in terms of modern physics, esoPhysics? Here, from the point of view of esoPhysics, Dr. Bartlett, applying one finger to a diseased point and another finger to another point, already healthy, performed quantum entanglement of these points, and then, as a result of a spontaneous quantum jump, the quantum states of these points were exchanged for opposite ones. That is, the state-sickness of this sick point, changed to the state-health of this sick point as well. The sick point thus took over the health from this healthy point. This is the natural quantum process of two quantum entangled entities, which involves the spontaneous exchange between these points of the quantum states of these entities. In quantum physics, as an example, two electrons are entangled with each other because of the spin of these electrons, (spin is also a kind of function of the state of an elementary particle). It is important that in such a jump the total quantum number of the process cannot change. In the example with electrons, if we have quantum entangled, due to their spins, two electrons, and one has spin (½) and the other has spin(-½) (in units of Planck's constant), so the total spin of the system of two electrons in this case is ½ +(-½) = 0. As I wrote in quantum entanglement there is a spontaneous jump, but the quantum number must be conserved, so after the jump the electrons must, I emphasize in this case must, exchange spin values with each other, so that there is a conserved $(-½) + ½ = 0$.

Quantum entanglement itself is due to a locality at the Unmeasurable Level and because wave physics applies at this level. What does wave physics mean? It doesn't mean that once entities are particles (corpuscles) and another time they are waves, as it is rather trivially used to be colloquially referred to. It only means that the physics of these entities, particles, field quanta, or in other words: how these entities behave, is wave physics. More precisely, if we have the Schrödinger wave equation, then the solution of this equation is precisely the complex wave function, which describes the physics of the whole system. - Wait, wait, after all, the wave function describes the physics of the whole system of particles, not of a single electron(entity) - you are about to tell me. I answer both yes and no. After all, this wave function of the whole system by means of the Fourier transform can be broken down into component functions for each (if we have, for example, an array of electrons) individual electron. And then the superposition after all electrons will give us the initial state function. And, interestingly, any such state function for an individual electron will also be a wave function. This can be done for any system of arbitrary particles. Thus, it is straightforward to see that every elementary particle, and even more: everybody, has wave physics at the Unmeasured Level, although it is not itself a wave. This applies even to light. Light is also a collection of particles, specific massless particles, quanta of the Electromagnetic Field, the so-called photons. It's just that the physics of light is wave physics. Light is distinguished to the extent that its physics from the Unmeasurable Level is wave physics and physics from the Measurable Level is wave physics. Thus, the dispute that has been going on since Young's famous experiment with the scattering of light on two slits, which supposedly confirmed the wave nature of light, esoPhysics resolves in favor of the supporters of Newton, who considered light to be a collection of corpuscles (particles). Thus, light is a collection of photons, particles, only that its physics is wave physics.

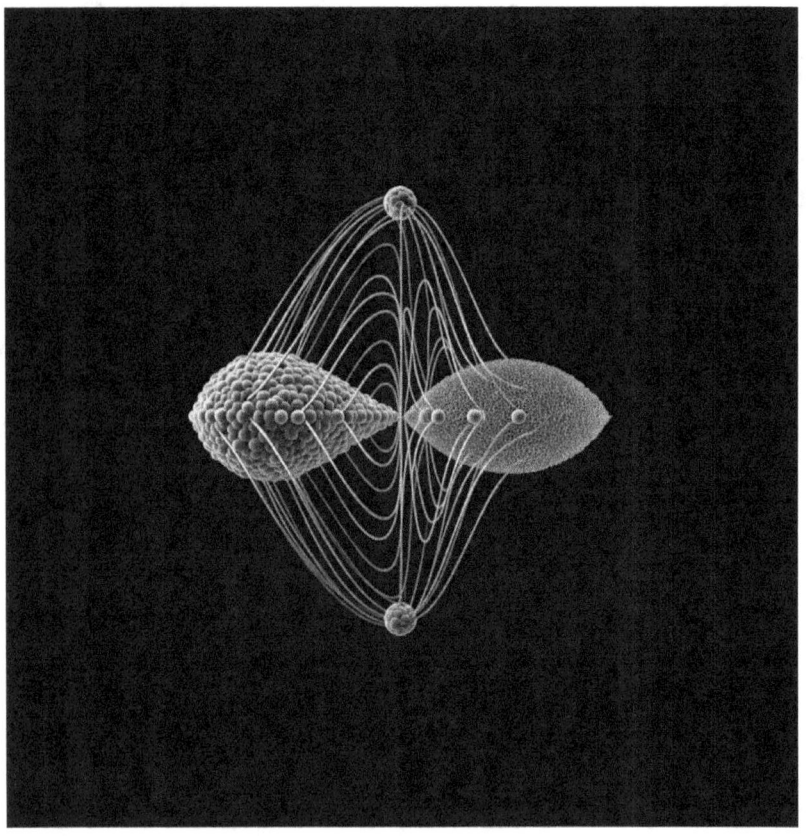

Light, as I wrote, is peculiar, because it is massless, hence it reaches absolute velocity, which is the highest that entities on the

Measurable Level can reach. This is a tremendous, but nevertheless finite, speed "c". However, all the time in the consciousness of some physicists lingers the view, as an aftermath of the fact that supposedly light is a wave, that it is correct to treat particles once as particles and the other time as waves depending on the experience in which these entities participate. And that this is what corpuscular-wave dualism supposedly consists of in. This is an erroneous position. Because, as I wrote, this corpuscular-wave dualism consists in the fact that at the Unmeasurable Level the entities (these particles) are described by wave physics, and at the Measurable Level the particles (these entities) are described by classical physics.

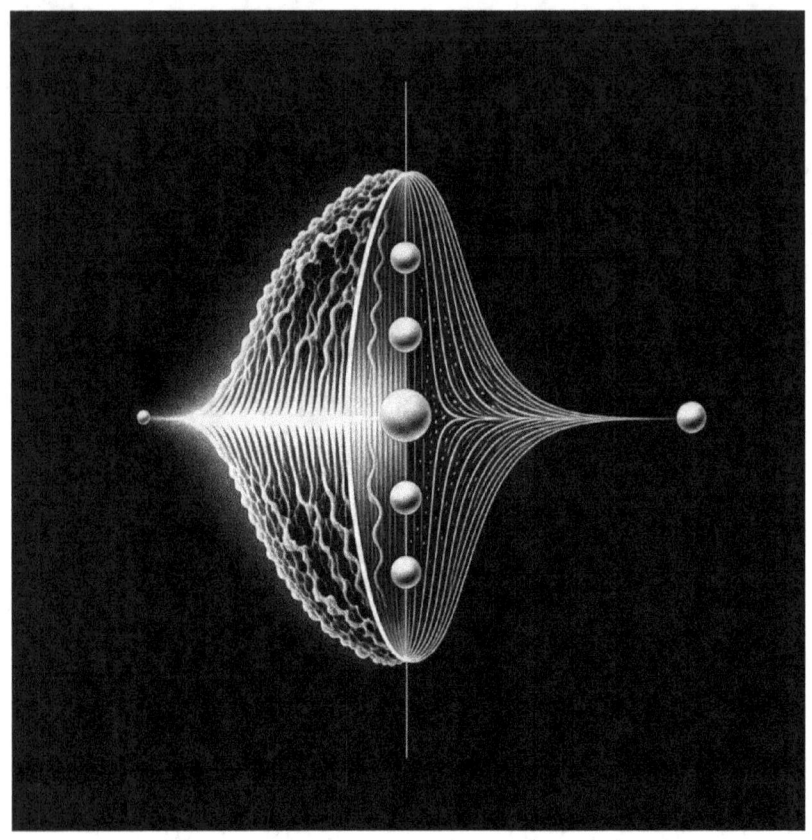

In summary, for these two reasons. Let me remind you, from the fact that the Unmeasured Level is a local, and from the fact that the physics of entities at this level is wave physics, and still from the principle of conservation of quantum numbers arises a very important issue. This issue that with quantum entanglement of two particles due to their state functions must, I emphasize must occur spontaneous quantum jumps, when these entities suddenly and unexpectedly exchange, they must so exchange, their quantum entangled state functions alternately. Let me remind you that quantum entanglement is the entanglement of two entities(particles) from the Measurable Level due to their specific state functions (let's emphasize quantum state functions) and the transfer of them (these two particles) to the Non-Measurable Level, where such quantum processes take place there (hence the name: quantum entanglement) that it results in the fact that at the Measurable Level there are constant jumps, spontaneous jumps, quantum jumps between these particles. That is, observers from the Measurable Level observe such jumps. And these are jumps of these state functions, these state functions of particles relative to which these particles are quantum entangled. We then speak of the coherence of quantum entanglement. And then processes are carried out at the Non-Measurable Level, the effects of which we then observe at the Measurable Level.

What are the processes at this Unmeasurable Level? Well, not too much is known, because there is an impassable barrier between these levels, between the Immeasurable Level and the Measurable Level. We flesh-and-blood humans are on this one border of these levels. At least in mortal life. However, from the mathematical formalisms of Quantum we can deduce what is happening there. And that's what physics is.

If we are dealing with electrons that are quantum entangled due to their spins, then here is the situation: we have state functions of these electrons, which boil down to the spins of these electrons; there are spontaneous jumps between these electrons, but the overall quantum number must then always be conserved; so if one electron had a spin ($½$) and another had a spin ($-½$), then between these electrons the entanglement requires alternating exchanges, jumps, of the spins of these electrons, so there was a conserved overall spin of these electrons, which is "0". Characteristically in the case of electrons and their spins, these state functions (these spins) are discrete functions, that is, they have only two values [½;-½], but in fact entanglement can occur between entities due to continuous state functions. And this is where it gets much more interesting. And this is where the field of mental Quantum Tools comes into play. Why? Because let's look at if we have two entities (particles) quantum entangled due to their continuous state functions. One entity A has a state function (W+) and the other entity B has a state function (W-), and now we will perform quantum entanglement of these two entities. This according to the theory, which is - confirmed empirically, after some time a quantum jump must take place between A and B, but so that the overall quantum number is preserved, which is (W+) + (W-) = 0 (this is how we assumed, I emphasize this is how we assumed, but this is not a necessary condition), but there is one trick here, these state functions are not discrete but are continuous. And what will happen after the first jump? Here B will take over (W+) from A, but because (W+) is not discrete, it will be a slightly smaller value of (W+)', and this value will be slightly smaller than (W+) and (W+)'<(W+), since so A must take over such (W-)' that (W-)'> (W-), because (W+)' + (W-)' = 0. After "n" jumps the situation will stabilize and the final function of the state of being of A will be (W+)' and it will be equal to about (W ½ +), and the point B will reach adequately the function of the

(W ½-) and it will be satisfied that (W +½) + (W -½) = 0

Well? -you ask.

Well, it is that if you manage to quantum entangle a problem, adequate with a continuous function (-), with the solution of this problem (+). It can be seen immediately that a simple way this leads, this quantum entanglement, to halve the problem (½-) during this, let's call it so, one session of quantum entanglement. And these are all quantum processes that are accomplished by changing at the Unmeasurable Level the Source Causes behind, in this case, the problem. Note that this applies to any problem, and quantum entanglement so used changes the Source Causes concerning that very problem. So, it is indeed a miraculous Philosopher's Stone that changes so, without physical work at the Measurable Level, the Source Causes pertaining to that problem, so that there is no longer a problem. Maybe not in the first séance, because then we can get rid of the maximum of the problem by half (½-), but after some time, say the next day again we can quantum entangle already (½-) half of the problem and we will get a maximum reduction by half of this half of the problem (0.25-). And so proceeding in a certain sequence of séances one can get rid of the whole problem. According to the sequence: ½;¼;1/8;1/16;1/32;1/64;etc. Amazing!!!

Only, does it make sense? How can this be accomplished?

Well, it has and we, every human being endowed with Free Will, can do it. The expression of man's Free Will at the Measurable Level is the work done at that level. And what does it mean that we can do work? It means that with this Free Will we can assemble elemental forces. But, in order to rearrange the elementary forces, one must also use force. So, this force is the Free Will of man. But, as I wrote, the Absolute has given us, humans, such an attribute that we can use Free Will also on the Unmeasured Level. How to do this, I will describe shortly. Now let's go to a small summary. Free Will is an additional Elemental Force, a side of the known four main forces (Gravity, nuclear strong and weak, and electromagnetic force). This force, this Causal Cause, can act both on the Measurable Level and on the Non-Measurable Level. On the Measurable Level, thanks to Free Will, we do the work in question, in order to achieve certain effects. So how do we use the Free Will in us to get certain effects through the Non-Measurable Level? The answer is simple: we need to focus our attention skillfully and quantum entangle two state functions in such a way, to "set them up" first, so that the result of this quantum entanglement is to change the source Causes of interest from the Non-Measurable Level, so that it results in the effect we want at the Measurable Level. It turns out that such actions are feasible. In the past it was called Magic. Today, physics, or rather esoPhysics is Magic. I can already hear those voices of indignation and opposition. Those words: - don't make it up, these are your confabulations. I answer: -in what I write and what I have included in esoPhysics there is not a shred of falsity. Wasn't it Murray Gall Man himself who stated: in quantum physics all those actions are allowed that do not break quantum laws?

The expression of the use of Human Free Will as an Elemental Force and Causal Cause and the use of quantum entanglement are the mental Quantum Tools I have formulated and developed, such as the Philosopher's Stone Algorithm and the directly resulting Cognitive Prosthesis. I dedicate these tools to Sir Isaac Newton. And what they "look like" and what they consist of I will describe shortly. But before that, I will return to the example of Resonance of healing music on a person. Let me remind you what I have already described in this book, we make a quantum entanglement of some of our problems (most often some disease) with an adequate, with appropriate harmonics music or light signal. Let's first note that both sound and light have wave physics, so they naturally lend themselves to quantum entanglement. Let's note that in this case the state function of the solution of the problem does not lose its value and has it all the time, after each quantum jump the same, so the problem itself already for the first session of such entanglement is significantly reduced, at least theoretically, as much as (1/n-). What this factor of 1/n will be is already theoretically difficult to determine, but it can be determined practically. It will depend on the "problem", the content and quality of harmonics in this signal, and on the length of the very seance of such a Resonance. Returning to the mental Quantum Tools, it can be said that my proposal, my quantum tools including the Philosopher's Stone Algorithm and the Cognitive Prosthesis are, as of today, the best and most perfect tools, it is the top of the top, it is probably the most perfect tool invented in the history of mankind, a tool which, just as pebbles used to be that proxy of humanity, so mental quantum tools are the proxy of the new species Homo Sapiens Quantum. Those who once try quantum tools, mental Quantum Tools, will probably use them for the rest of their lives. They allow to change the root causal and intentional causes behind given effects, behind given problems, which manifest at our Measurable level in our reality, in our life on earth, at our level of matter. They allow us to model reality, they allow us to change problems radically. All we need to do is to use quantum tools or quantum laws, quantum entanglement, and entangle skillfully a problem, a given problem, with the solution to that problem. I repeat, it is not acting on the problem itself, it is acting on the causal and intentional causes that stand at the unmeasurable level behind these problems, which then result in these problems precisely at the Measurable level. This is probably a greater achievement than the invention of electromagnetism, because man gets to have at his disposal, in this scientific sense, his own free will as an elemental force, a causal cause, and he can use this free will at the Measurable level to correct the root causal and intentional causes at the Non-Measurable level. Let's move on to the mental Quantum Tool itself in this case the philosopher's stone. We must use to the maximum the elemental force that is our will. We choose as an entity, as a point, some finger, of our, let's say, right, hand this point identified with the problem. We can do this because it is Free Will that allows us to create reality in this way. So we picked point No. 1, this is the thumb associated with the right hand, this is the point associated with the minus. To this point we assign a given problem, a specific one, a certain one. With our free will, we place this point with this problem at the Measurable and Non-Measurable level, then we select the thumb in our left hand as the point associated with plus, as the state desired by me, by us. To this point we will assign with our free will the solution to this problem from the first point, of course, as you know, here it is not about getting a solution to the problem in some miraculous way only to influence the causal and intentional causes behind the problem so as to change them in favor, so as to cause the problem to be solved. We will attribute to this point number two just this solution to this problem, and we place this point number two also at the non-measurable level. And now we come to the clou of this whole process, namely we make with our free will-we say it and we incant it in our mind-we now make a quantum entanglement of point number one with point number two. We place this quantum entanglement sacred in the space of our heart, where there this entanglement will maintain the coherence of this quantum entanglement in a permanent way all the time, that is, all the time the processes will happen on an unmeasurable level even without our consciousness, because these processes do not depend in any way on consciousness, these are quantum physical processes, they do not depend in any way on consciousness. An entanglement has been made, and this entanglement will now persist at an unmeasurable level and work quantum all the time. In this state we can last for tens of minutes, even de facto at the same time the whole process can be compared to some kind of meditation, because we are then actually limited to mindfulness, to observing our own body, to observing our own breath, to being aware of our own breath. And so we continue. At some point we may receive, after some time, clear signals from the body, some sort of, like, yawning, shuddering, shaking, spasms of the body, and this is the information that this process should be completed. We do this in such a way that we incant that by the power of our free will we now perform a decoherence of quantum entanglement between point 1 and point 2, if in doing so we give off negative energy that could harm us, other people harm us, we ground this energy. Why this is important is to ground such energy, because in the course of quantum processes, if we heal something, de facto then other quantum numbers are affected, and since there is a general principle of quantum conservation, if it improves somewhere, it must deteriorate somewhere, and this is the negative energy that can flow out somewhere. In a word, we entanglement creators can start harming ourselves and other people. This is negative, so it is worth it to prophylactically ground such energy already for our own safety and for the safety of people around us. This seems perhaps a bit like witchcraft now, but for centuries that's what psychics have also done, is that they purified themselves. Anyway, as I write myself, magic is otherwise an intuitive application of quantum laws, and they simply grounded prophylactically all those negative energies that were being released in the quantum processes they were doing. This has a deep justification precisely on the basis of Quantum, on the basis of the behavior of quantum numbers. So, as you can see, the application of quantum entanglement, the setting of state functions of entangled willing entities has a magical connotation, it is similar more to magic, so from this it follows that in this way physics and practical application has a magical, esoteric connotation. It can also be seen that with this, Newton's dream of making physics more like magic was fulfilled. Newton in his youth developed the mathematical formalisms of classical physics, Newtonian dynamics, and then in later years in such a magical way he wanted to tease out the secrets of this world to nature. This he failed to do for obvious reasons. It now means that ezoPhysics and its practical application are a significant nod to Newton, to those 18th century alchemists who formed the foundations of chemistry, who created the teeth of physics.

I am the creator of the mental Quantum Tools, such as the Philosopher's Stone Algorithm and all other tools resulting from this very Algorithm. These are: Cognitive Prosthesis or

Elixir of Life and others, but these are simply detailed and more selective applications of this Philosopher's Stone Algorithm. Let's focus on the Philosopher's Stone Algorithm itself to try to understand in depth the essence of its operation. The source of this quantum tool is to "cure" and solve the problems that happen to us in life. Whether they be health problems, emotional problems, social problems, psychological problems or some other. It applies to all problems in general. In this tool, we set up the problem and the solution to the problem in such a way that, with the help of quantum entanglement, we get to the root Causal and Purposeful Causes and these problems and change them effectively. It is through this screening of the Philosopher's Stone Algorithm.

Algorithm Philosopher's Stone - Proper procedure

As I mentioned in the procedure Algorithm Philosopher's Stone converges the application of several Quantum Laws. First, we use our Free Will, and it is used extensively. As an Elementary Force, which is used to create the two points that we will ultimately quantum entangle, to determine and set the state functions for these points through which we will perform the entanglement, and to perform the single seance itself. Surely the most important thing in all this is the act of quantum entanglement itself, which we also perform by expressing (in thought or verbally) Free Will. Finally, we also express in thought or verbally the act of breaking this quantum entanglement (the act of decoherence). We also make sure with our Free Will that all the garbage that appears in the quantum processes accompanying this seance is digested into the "trash" i.e. literally grounded. This is all very important, because you need to maintain a kind of health and safety hygiene throughout this procedure. For your safety as well as for the safety of other people. According to the Quantum Laws, quantum numbers must be preserved. So if we "fix" something quantum, the corresponding quantum numbers will change, but since quantum numbers are immutable, another "problem" will pop up somewhere if we don't ground the "dirt" in the right way. If you forget about it, your efforts will be of no use, for Nature will claim its own. You may even, for example, be cured of an illness, but immediately afterwards some other equally troublesome or even worse ailment will take control of your body. Fortunately, the Earth is powerful, the Earth will endure a lot, so ground all the dirt, dirty energies calmly. I know this, in a way, from my experience, because as a forerunner of this method I went through this on my own. And I remember it painfully. Fortunately, later modeling of esotericists working with pendulums helped me get to the bottom of it. It is, by the way, an old and intuitive law of Magic that one should carefully purify (ground) oneself while "conjuring". As I mentioned, quantum entanglement is done between two specific entities because of the property that gets entangled. That is, using Quantum Tools, such entities must be created, created. This can be done mentally, and the empirical evidence for this is that this is what people, those who apply these Quantum Tools, commonly do. This is the case with the Two-Point Method, this is also the case with the Philosopher's Stone Algorithm. And, as you know, empirical proof in physics (esoPhysics) is conclusive. In general, what I write in my publications, and in this publication too, is confirmed empirically, if only by my personal practices, but also by the practices of the developers of these methods in many European countries and in the USA, and all over the world. So, they have, these theses, almost Scientific confirmation. I write almost, because the official mainstream of Science has a shtick of interpretation of single-level Quantities and a typically materialistic and atheistic paradigm, and such Science does not actually recognize similar content. As I mentioned such Science, however, leads to contradictions and numerous paradoxes, which is worth remembering. And only the Theistic Interpretation of Quantum, on which all these theses I write in this publication are based, do not lead to contradictions. As you can see, these theses are also empirically confirmed. But, as you know, there are many things in science that are empirically confirmed, but the official mainstream of Science does not accept this. So as not to be lip service I will give the example of Radiesthesia, which is persistently, despite the overwhelming empirical evidence, ignored by Science.

In mentally creating the entities we intend to quantum entangle, we use the creative power of our Free Will. So, if we already have two concrete entities, we still have to with our Free Will adequately determine, create state functions (also using Free Will) for entity one, which will represent the problem, and for entity two, a function that will represent the solution to the problem. Note that in such a case (problem)+(solution of the problem) =0, i.e. the main quantum number for this quantum entanglement will be zero (0), which is decisive for this entire physical process, because this value, despite numerous possible quanta jumps, must be preserved.

So now I proceed to fully describe the scheme of the Philosopher's Stone Algorithm: Sit somewhere in a comfortable posture or lie on your back. Try to force your family not to disturb you during this time (a good few tens of minutes). You can be accompanied by quiet relaxing music (Chillout music) in the background.

[Incant (enchant?) in thought or whisper].

...I peel off point No.1, this is the point associated with my right thumb, this is the point associated with (-), this is the actual state that exists.

By the power of My Free Will from My Transcendence....

...I assign to point no.1 a state function that defines and expresses ...

(and here you weave in your problem, your illness of soul or body, or any other kind of life problem)

   ...??

  ...

... By the Power of My Free Will from My Transcendence, I place Item No.1 on the Measurable Level and the Non-Measurable Level....

[This is the first step of the Algorithm].

Another second step of the Algorithm:

[Incant].

...I peel off point No.2, this is the point associated with my left thumb, this is the point associated with (+), this is the state I desire.

By the power of My Free Will from My Transcendence, I assign pointwise No.2 to the function of the state, which is determined by the Point in the Divine Matrix (Akasha Chronicle) of Energy (Healing), which means....

???

...

[Incant].

... this Point in this Divine Energy Matrix (Akasha Chronicle) chosen by me determines the solution (+) of this problem (of Point No.1), determines the function of the state, which I by the Power of My Free Will assign to Point No.2.

By the power of My Free Will, I place Item No.2 on the Measurable Level and the Non-Measurable Level....

[This was the second step of the Algorithm].

The next, third, step of the Algorithm:

[Incant].

... By the Power of My Free Will I am now making a quantum entanglement of point No.1 with point No.2 due to the state functions of these points.

By the power of My Free Will, I place such quantum entangled points No.1 and No.2 in the Sacred Space of My Heart in the Field of My Heart....

...let There, with the coherence of this quantum entanglement preserved at all times, permanent and positive quantum processes occur for me, consistent with the intention of this entanglement. If there secretes in an additional way negative energy that could harm me or others, by the Power of my Free Will, I ground it in a safe way....

(A period of passive waiting follows, which can last quite a long time. However, if there is a sudden breakthrough, perceived as some kind of spasm, shivering, vibration, excessive yawning or other clear body signals we end this phase and go straight to the last step of the Algorithm).

The final step of the Algorithm:

[Incant].

...By the Power of My Free Will I am now performing a decoherence (breaking) of the quantum entanglement between points No.1 and No.2

If, in the process, I emit additional negative energy that could harm me or others, by the Power of My Free Will, I ground it.

(After completing the Algorithm, you shake your hands towards the ground and clap to end the session).

As can be seen in this diagram, as entities that are subject to quantum entanglement one peels any of the fingers of the left and right hands, most often they will be thumbs, and then sets (formulates) state functions for these points. The Philosopher's Stone Algorithm, constructed in this way, allows much more than a selective, that is, working on one problem, Two-Point.

Practically, the setting of such a Philosopher's Stone Algorithm can be very complex and cover many problems at once. This is an expression of this multitasking at the Non-Measurable Level, to which I have already referred many times, and on which, among other things, the principle of operation of the technical Quantum Tool that is the Quantum Computer is based. After that, you only need to adjust the equally complex state function for solving the problem and you can carry out an effective treatment. I will soon show with examples how to do this and what I mean, but you probably already guessed, dear reader, what the essence is. I often use very long settings of the Algorithm. Of course, for it to be properly effective, however, more time is needed for the screening itself. However, it can be compared to such a decent daily meditation. However, the Philosopher's Stone Algorithm is something much more valuable than ordinary passive meditation. It is active, it is effective, it allows you to model reality and health. Meditation merely calms the nerves and the work of the Central Nervous System. Yes, I agree, meditation is very valuable, and I recommend it to everyone, but the Algorithm Philosopher's Stone is much more valuable, it allows you to be active, to influence the modeling of reality. It is simply a Quantum Tool. At the formulation of the theoretical basis of the Two-Point Method in America, it was noted that the Two-Point is much more effective when "located" in the Heart Field. This was discovered by Dr. Kinslow, one of the forerunners, along with the creator Dr. Bartlett himself, of the Two-Point Method. It turns out that this is also true for the Philosopher's Stone Algorithm, and it is due to the fact that in the Heart Field it is easiest and most stable to obtain and maintain quantum entanglement coherence. That is, then it is the Quantum Tool that "works" at the Unmeasurable Level, Quantum-wise. Conversely, with decoherence, the Quantum Tool is "ejected" from the Non-Measurable Level and does not work Quantum. This is the power of Quantum Tools, that they make changes in Source Causes from the Non-Measurable Level. Which ultimately manifests in the obtained desired Effects at the Measurable Level. This is a real phenomenon, it is as if to a large 3D Printer to give on the input a picture of the expected thing and so completely "without getting your hands dirty" to get on the output just that product. The only condition is in the case of these Tools that Quantum Numbers must be preserved.

It must be honestly admitted here, some may be put off by this, such an almost overt reference to magic. But let's say it openly, all activities that will use Free Will as an Elemental Force, as a Causal Cause, will have a problem with it. This has not been the case so far, with the exception of just Magic, hence such associations. Perhaps it is time in Science for another Copernican Coup. If we have begun with the use of mental Quantum Tools a new era, the era of Quantum Man, then Science should also be reformed thereby.

Since this is an item dealing mainly with quantum psychology, we will limit ourselves here primarily to the Cognitive Protector, a clone of the Philosopher's Stone Algorithm limited to getting the mind in proper shape and acquiring health in the broadest sense.

Formally, the Cognitive Prothesis is no different from the Philosopher's Stone Algorithm except for the "setting" of the problem and the solution to that problem. It is this "setting" of the problem and solution of the problem that makes the Cognitive Prothesis unique. This "setting" is another way of formulating the problem and formulating the solution to the problem. It makes use of this multitasking, and one can describe such a generalized problem in a truly multithreaded way. This I will later present with examples. Personally, I myself sometimes "set" such a Cognitive Prothesis sometimes even more than half an hour. This is not a joke. In fact, this notorious multitasking works and such a Prothesis or Algorithm in general works on many things (sub-problems) at once, at the same time, just like a quantum processor works. However, one must always remember, and I say this purely from empiricism (from my own empiricism), to always carefully ground oneself from unwanted energies that can always "flow out" somewhere, which can even be dangerous for the user of such mental Quantum Tools. And in order not to fall from the rain into the gutter, remember to always guard against this, even if it were to prolong such a seance itself. Also remember, for the mental Tools to be as effective as possible, a whole series of séances should be conducted, and not limited to one as in the Two-Point Method. The Two-Point Method is merely an imperfect prototype for the mental Quantum Tools I have formulated, from which its many shortcomings and theoretical gaps arise. But, of course, all the time remember that without the Two-Point Method there would be no my mental Quantum Tools. It is puzzling to me that the Two-Point Method was discovered almost a full century after the discovery and formulation of Quantum Mechanics, and that this Two-Point Method was discovered quite by accident. Probably no one would have paid any attention to this fact, if it were not for the fact that the Two-Point Method itself proved to be quite effective and a fashion for it took hold, at least in Europe. I, too, came across this Two-Point completely by accident, and at first, I even disregarded it and dismissed it as an invention and phantasmagoria. But later, however, it came to me in some flash of inspiration how valuable this discovery was. On the basis of this Two-Point Method, it came to pass that I developed an entire esoPhysics and mental Quantum Tools of my own authorship. Mental Quantum Tools is an expression of the practical application of esoPhysics. After reading my books such as: "Cognitive Prosthesis", "Quantum Conditioning of the Mind", "Four Pillars" and others, anyone can actively deal with modeling their own reality, their own mental and physical health, their own mental condition. For many of the problems of this sphere, there has simply been no effective "cure" so far. I mean, come back, people always discovered something there, they always had some remedies for problems, but there was no such good, such effective remedy for problems. It is only the mental Quantum Tools that make it possible, perhaps for the first time, to face problems that until now were practically "incurable" and often provided fodder for primitive pagan practices of treating such cases as sacrifices to the cruel "idols" of Nature. Unfortunately, even in the 21st century there are people who cultivate such pagan customs, and these people, horror of horrors, are completely unpunished. Now, however, thanks to mental Quantum Tools, people will have, at least minimally, a shield from such pagan aggressions.

6. practical application of mental Quantum Tools

In the practical part, I will actually focus on how to construct "settings" of mental Quantum Tools. Such as the Philosopher's Stone Algorithm and its clone the Cognitive Prosthesis. I will admit that I will refer here in part to the books I have already written. Whether a given Algorithm or Prosthesis, in this particular case, will work well or worse at the Non-Measurable Level is determined, at least from my practice so, by whether we make the setting of this tool with sufficient precision, in a word, whether we have bitten into the essence of the given problem. For let's consider such a problem: I have diabetes. It would seem that the natural solution to the problem here would be: I do not have diabetes.

Well, this setting of the Algorithm may not be sufficient. It would be necessary to make the problem more specific. For example: I suffer from insulin resistance or have destroyed islets of Lange Hans, in addition, I am obese, have atherosclerosis and have symptoms of psychosomatic disease. In this case, on the other hand, in the solution of the disease would be: I do not suffer from insulin resistance or have healthy islets of Lange Hans, not destroyed, revitalized, my weight normalizes, atherosclerosis goes away and I do not suffer from any psychosomatic diseases. I feel great, my sugar is normal. In this case, the mental Quantum Tool will penetrate deeper into the Source Causes of the problem and change these causes in the desired way. Such a setting of the Philosopher's Stone Algorithm will be more effective than such a simple formula that: I have diabetes, and that: I do not have diabetes. It would seem that this is a simple and insignificant difference, but, as I wrote, in the case of the application of Free Will in physical processes, it is nevertheless significant. Sometimes one word, one meaning can have a decisive impact on the effectiveness or ineffectiveness of a given mental Quantum Tool.

a. Psychological prevention

Let's now turn to practical examples of the settings of various Cognitive Prostheses. Let's assume that we have this case: we are in our fifties, for simplicity's sake let's assume that we have the case of a man, although it applies equally well to women. So, I am in my fifties. I'm already tired of life, I'm in andropause. My capabilities are no longer what they were when I was young. I already know my limitations, and despite this, I can't accept the passing of life. I have a family, a wife, children. I am a respected businessman who has some property. I live a high standard of living, but nevertheless I feel burnt out, I don't want anything. My metabolism is gradually going down, I observe progressive weight gain in myself, etc.

What can be done here? Because, after all, I am neither sick nor infirm, nor am I old. Old age is not yet, but I am burnt out and discouraged about life.

Let's use the appropriate Cognitive Prosthesis with the given setting.

Sit in a comfortable posture or lie on your back. Warn your family not to disturb you.

[Incant (enchant?) in thought or whisper].

...I peel off point No.1, this is the point associated with my right thumb, this is the point associated with (-), this is the actual state that exists.

By the power of My Free Will from My Transcendence....

...I assign to point no.1 a state function that defines and expresses ...

I feel mentally weary, emotionally tired. I have overloaded in my brain: emotional system, emotional system, logical thinking system, sense of security system. My dopamine systems in my brain, the reward and control systems are dysfunctional. There is abnormal secretion of dopamine and serotonin, as well as norepinephrine in my brain. In addition, cortisol is also excessively secreted. The normal relationship between cortisol and the neurotransmitter GABA is disrupted. I become overly stressed by my problems, which only compounds their harm to me and my body. I am on the verge of mental exhaustion. The system in my brain: Prefrontal lobes - Frontal lobes - Amygdala - Hippocampus, is overloaded, malfunctions, is overused. All these negative processes build up and progress in an unfavorable direction for me.

...

... By the Power of My Free Will from My Transcendence, I place Item No.1 on the

Measurable Level and the Non-Measurable Level....

[This is the first step of the Algorithm].

Another second step of the Algorithm:

[Incant].

...I peel off point No.2, this is the point associated with my left thumb, this is the point associated with (+), this is the state I desire.

By the power of My Free Will from My Transcendence, I assign pointwise No.2 to the state function, which is determined by the Point in the Divine Energy Matrix (Healing), which means....

I feel like a newborn. I feel healthy, rested mentally and emotionally. My emotional system in my brain is strong, healthy, rested, powerful and balanced, cleansed. My emotional system is healthy, strong, rested, powerful and balanced. My logic system is strong, powerful and healthy. My feeling system is strong, powerful and strong. My dopamine systems in the brain: the reward and control systems are healthy, efficient, without distortion. Dopamine and serotonin are secreted in the right healthy proportions. Norepinephrine is secreted properly and healthily. Cortisol is not secreted in excess in my brain and body. The neurotransmitter GABA is secreted alongside Cortisol in proper healthy proportions and relationships. My problems no longer stress me out, I solve these problems on the fly. I already feel good, in full control of myself, my emotions and my psyche. The system in my brain: Prefrontal lobes - Frontal lobes - Amygdala- Hippocampus works well, in an optimal and healthy way for me. Life is not overwhelming me. I have good and healthy interpersonal relationships with others, I have good and healthy social behavior. I feel excellent, I am mentally and emotionally healthy. This is a positive and lasting change in me and the world towards me.

...

[Incant].

... this Point in this Divine Energy Matrix chosen by me determines the solution (+) to this problem (of Point No.1), determines the state function that I by the Power of My Free Will assign to Point No.2.

By the power of My Free Will, I place Item No.2 on the Measurable Level and the Non-Measurable Level....

[This was the second step of the Algorithm].

The next, third, step of the Algorithm:

[Incant].

... By the Power of My Free Will, I am now performing quantum entanglement of point No.1 with point No.2 due to the state functions of these points.

By the power of My Free Will, I place such quantum entangled points No.1 and No.2 in the Sacred Space of My Heart in the Field of My Heart....

...let There, with the coherence of this quantum entanglement preserved at all times, permanent and positive quantum processes occur for me, consistent with the intention of this entanglement. If there emits in an additional way negative energy that could harm me or others, by the Power of my Free Will I ground it in a safe way....

(A period of passive waiting follows, which can last quite a long time. However, if there is a sudden breakthrough, perceived as some kind of spasm, chills, vibration or other clear body signals we end this phase and go straight to the last step of the Algorithm).

The final step of the Algorithm:

[Incant].

...By the Power of My Free Will I am now performing a decoherence (breaking) of the quantum entanglement between points No.1 and No.2

If, in the process, I emit additional negative energy that could harm me or others, by the Power of My Free Will I ground it.

(After completing the Algorithm, you shake your hands towards the ground and clap to end the session).

Such a session should optimally last about 1.5 hours. There may initially be negative feelings and apparent deterioration of mood. So one should also be careful with this. If someone will feel it too "painful", I suggest then to shorten the one-time seance by half, and if even this will not be enough then shorten it even more. And then progressively increase it only. This is because during its use and afterwards, all the traumas, irritations and unpleasant sensations that we have experienced, that we have accumulated over several decades of life, are released. So, it would be best to wait it out during the screening. After all, they didn't kill us "then", so they shouldn't do anything to us now during their purification. However, if one has fears and aches of these renewed, released psychic "pains", he can measure this Cognitive Prosthesis. Ultimately, this Cognitive Prosthesis, of course, the cycle of such séances, restores mental and emotional efficiency, the potential of the mind. It gives the effect of subtracting several decades of life so godly, such a life from day to day. Restores youthful lightness and clarity of mind. I will propose similar settings, but in slightly different configurations, of the Cognitive Prosthesis later in this publication.

Being in the part of the book that does not touch on specific ailments, but is concerned with a certain prevention of life, let me now present the Cognitive Prosthesis so "set" that I gave it the name Elixir of Life, as it concerns the implementation of long-term action to enjoy as long as possible a good life and good condition of body and spirit.

Elixir of Life

The quality of life is basically determined by the efficiency of our vegetative system. Actually, the quality of human life is determined by the autonomic and vegetative nervous system. Mainly the sympathetic and parasympathetic nervous system. As the name suggests, paraphrasing it, whether we will have a good, trouble-free life or, conversely, a host of problems resulting from the "bad" work of our bodies, bodily shells, is determined by these nerves. In other words, whether we will have a reasonably pleasant or unsympathetic life is mainly decided by the sympathetic and parasympathetic nervous systems, their proper and optimal work or their dysfunctions. Someone will immediately here shrewdly point out: what about the brain, intelligence, higher feelings? Yes, they are also important, even very important, but the physical body has its own laws, which were developed by Nature much earlier than intelligence and mind. Anyway, in fact, these cannot be separated. I am about to propose a kind of Philosopher's Stone Algorithm, which can be confidently called the Elixir of Life, because it will deal with the health and physical condition of the human body, and will increase our life potential. In general, the Philosopher's Stone Algorithm has a more general effect and transforms any type of problem into its solution. Here, however, it will be concerned with a person's health and maximum long life, in the broadest sense of the word. Health, as I have already written, is affected by the state of our autonomic and vegetative nerves, but also our mental and physical condition is affected by some body defects that we have had for many years, chronic diseases, also many years, stresses and behavioral experiences, actually negative behavioral behaviors that impair our psyches and characters, which also affect the condition of the body. All this I will try to include in this Algorithm. Returning to the vegetative system, it can be said that all sorts of defects in this area. That is, blockages, minor, major paralysis and even degeneration of these nerves lead to a number of diseases and dysfunctions, inflammations and diseases of internal organs, which ultimately lead to pathological changes of internal organs and organs. The solar plexus and the vagus nerve are particularly important here. And here all blockages or paralysis have a destructive effect on the quality of a person's life. For example, blockages in the solar plexus and vegetative innervation can also cause chronic halitosis (bad breath). If the dysfunction of the solar plexus and vegetative innervation is permanent, such a problem (halitosis) can be persistent. However, it is reversible, if only by the skillful application of Quantum Tools (for example, the Philosopher's Stone Algorithm). As the name suggests, the solar plexus is such a Sun of our body, the most energetic nerve. From there, energy radiates to a multitude of organs and organs in our bodies. On the condition of the solar plexus depends the functioning of the entire abdominal sphere, the digestive system, including the condition of our intestines, on its condition depends the health or illness of the organs and organs of this area of the body. Occultists also believe that the solar plexus is one of the most important buses of the subconscious.

The legitimate question is: and what can cause dysfunction of the vegetative innervation, including the solar plexus or vagus nerve? The answer is simple: mainly stresses and emotional experiences, well, just life itself. After all, humans are psychosomatic beings. So if we know what so strongly affects the quality of our lives, let me remind you: old untreated diseases, some disability, the state of autonomic and vegetative innervation, unfavorable behavioral changes, our pernicious addictions and habits, overweight (underweight?), then what can we do? We can apply the Philosopher's Stone Algorithm. And right now I will give you, the reader, such a suggestion. Of course, you don't have to literally monkey with it, you can make the appropriate modification so that this scheme fits into your needs.

You can perform the algorithm for yourself or for any person. However, you must remember that by performing it for someone, you are karmically binding yourself to that person, and this may already bring unexpected and difficult to predict consequences in your life and his.

[Incant (enchant?) in thought or whisper].

...I peel off point No.1, this is the point associated with my right thumb, this is the point associated with (-), this is the actual state that exists.

By the power of My Free Will from My Transcendence....

...I assign to point no.1 a state function that defines and expresses ...

As of today, my physical and mental condition is defined. I have lived through several decades, during which I acquired several ailments of a pathological nature that plague me to this day [????? <u>Describe your problems</u>]. I have some dysfunctions of vegetative and autonomic innervation in the nature of blocks, paralysis of these nerves and even degeneration. My interpersonal relations with other people are moderately correct, but the state of my health, physical ailments and relations with family and people have led me over the years to numerous and unfavorable behavioral changes in my brain, mind and innervation, which negatively affect my mental condition, they have even imprinted negatively on my character and quite emphatically hinder my life. I also have a fatal addiction [???? <u>Name it</u>], I am overweight, I am tired of life, I feel that I have wasted my life. My base vibration level, the consciousness map, on the Hawkins scale is quite mediocre and does not exceed the level of Pride, which is 175 units on the Hawkins scale. All of this is having a disastrous effect on my psyche and on my physical and mental health.

...

... By the Power of My Free Will from My Transcendence, I place Item No.1 on the Measurable Level and the Non-Measurable Level....

[This is the first step of the Algorithm].

Another second step of the Algorithm:

[Incant].

...I peel off point No.2, this is the point associated with my left thumb, this is the point associated with (+), this is the state I desire.

By the power of My Free Will from My Transcendence, I assign pointwise No.2 to the state function, which is determined by the Point in the Divine Energy Matrix (Healing), which means....

I feel good physically and mentally. My brain and autonomic and vegetative innervation, including my solar plexus, vagus nerve and gastrocnemius nerves are undergoing neurogenesis of old and damaged neural and neuronal structures in the brain, consistent with the neuroplasticity of the brain and the entire nervous system. I get rid of all blockages, paralysis and degeneration in innervation. My internal organs and organs revive and get an extra positive kick. I feel younger and healthier every day. In my brain, the damaged neurons and brain structures are regenerating and recreating themselves in this direction, so that I become a true sigma male, a man who is favored by fate, so that I become an optimist in life, so that my character changes in this positive direction, have very good interpersonal relationships, so that I succeed financially and materially. May these changes in my brain and mind allow me to enjoy life, so that my partner and children respect and love me, and so that I can give them love. I am getting rid of all the unfavorable behavioral changes in my brain, mind and innervation that have blocked me so far and distorted my character through this pathway (through proper neurogenesis). I have strong and resilient systems: the mental system, the emotional system, the emotional system and the system of my sense of security. I am mentally and physically healthy. All my aging and pathological changes in my body [???? <u>Describe these changes</u>] are gradually receding, day by day. I am getting rid of old habits and addictions [???? <u>Describe them</u>], day by day. My body weight is correct and proper. My base vibration level (consciousness map) increases significantly and is at the level of Courage, that is, 200 units on the Hawkins scale, which leads to the fact that I am not afraid of life and the challenges that lie ahead of me. And all these are positive and lasting changes in me and the world towards me.

[Incant].

...this Point in this Divine Energy Matrix chosen by me determines the solution (+) of this problem (of Point No.1), determines the state function that I by the Power of My Free Will assign to Point No.2.

By the power of My Free Will, I place Item No.2 on the Measurable Level and the Non-Measurable Level....

[This was the second step of the Algorithm].

The next, third, step of the Algorithm:

[Incant].

... By the Power of My Free Will, I am now performing quantum entanglement of point No.1 with point No.2 due to the state functions of these points.

By the power of My Free Will, I place such quantum entangled points No.1 and No.2 in the Sacred Space of My Heart in the Field of My Heart....

...let There, with the coherence of this quantum entanglement preserved at all times, permanent and positive quantum processes occur for me, consistent with the intention of this entanglement. If there secretes in an additional way negative energy that could harm me or others, by the Power of my Free Will I ground it in a safe way....

(A period of passive waiting follows, which can last quite a long time. However, if there is a sudden breakthrough, perceived as some kind of spasm, chills, vibration or other clear body signals we end this phase and go straight to the last step of the Algorithm).

The final step of the Algorithm:

[Incant].

...By the Power of My Free Will I am now performing a decoherence (breaking) of the quantum entanglement between points No.1 and No.2

If, in the process, I emit additional negative energy that could harm me or others, by the Power of My Free Will I ground it.

(After completing the Algorithm, you shake your hands towards the ground and clap to end the session).

This is what my proposal for the Algorithm of the Philosopher's Stone, called by me the Elixir of Life, would look like. Of course, this is just a proposal. I'm sure you can modify this scheme and adjust it accordingly. Perhaps you have some unique problems that you should "cure" by this means in order to feel fully healthy and fulfilled. But overall the scheme is simple and universal. That's what this sculpting in the brain, mind and innervation, and in the body in a Quantum way (with Quantum Tools) is all about, fast and radical. It is possible to use this way to improve this ordinary Path of Spiritual Development of ours, which we are taking here clumsily at the Measurable Level. Importantly, these Quantum Tools can be used at any age.

These two suggestions for Cognitive Protein settings were merely concerned with the preventive treatment of one's own mental condition. However, it turns out that the possibilities of mental Quantum Tools are much greater. They can be used to correct practically all deviations regarding our mental well-being and mental health. By the way, I will mention that perhaps even more powerful from the point of view of treatment would be the use of Resonance of sound and vision signals, because, as I have already written in this book, they have the efficiency $(1/n\ -)$. That is, they weaken the problem much more strongly after one séance than the Philosopher's Stone Algorithm, which has an efficiency of only (½-) in this regard. But the Philosopher's Stone Algorithm or Cognitive Prosthesis has the advantage that it can be "set" for practically any problem, which cannot be done with Resonance, and also the Algorithm works on the principle of multitasking, and by appropriately repeating the Algorithm's séances anyway, the same effect is achieved as with Resonance.

From my experience, so to speak, as a precursor, I can say that if we carry out a séance of the Philosopher's Stone Algorithm with a length of, say, 1.5 hours, then automatically, well, perhaps after some time so proceeding, our effective sleep will decrease by $2*1.5=3$ hours. That is, we have an additional 1.5 hours of effectively free time. And people who need 8 hours will suddenly sleep 5 hours, or maybe even less. This is due to the fact that during the séance of the Philosopher's Stone Algorithm (Cognitive Prosthesis) similar physiological processes take place as during ordinary sleep. Sleep, in general, is a physiological time and process, during which the body performs a self-cleaning of all those processes that we experience so godly. This is a necessary period and process, but it turns out that by using the Philosopher's Stone Algorithm we, so to speak, take this sleep process out of such a function, hence the time saving later on during already real sleep. Only I would warn everyone to make sure that after such an abbreviated dream everyone can objectively and in good conscience use a car or other motor vehicle later. I would advise first to carefully see how these mental Quantum Tools will affect us in the long run.

b. Maps of consciousness, Relative Moral Energy Level µ.

In the context of human psychology and psychology, a very interesting idea is the so-called map of consciousness, or the level of the base (fundamental) Vibration that each person is, according to David Hawkins. This well-known American psychiatrist developed this idea in the second half of the 20th century and successfully spread his idea in the US and around the world. The idea is that the consciousness of each person is characterized by a certain coefficient of a logarithmic (???) nature reflecting the level of spirituality of a given person. Here's me more how such a table of consciousness map in Hawkins units represents.

David R. Hawkins' Levels of Consciousness Map depicts the different levels of consciousness, which can be divided into three main categories: lower, middle and higher levels. Here are the basic levels of base vibration according to Hawkins, along with examples:

Lower levels of consciousness:

1. **Shame** (20) - A person at this level may feel worthless and unworthy.

2. **Guilt** (30) - Feelings of guilt and self-recrimination predominate.

3. **Apathy** (50) - Lack of energy and motivation to act.

4. **Grief** (75) - Feelings of sadness and loss.

5. **Fear** (100) - Fear of the future and the unknown.

6. **Desire** (125) - Strong desires and needs that can lead to addictions.

7. **Anger** (150) - Anger and frustration.

8. **Pride** (175) - A sense of superiority and selfishness.

Medium levels of consciousness:

1. **Courage** (200) - Willingness to take on challenges and change.

2. **Desire** (310) - Enthusiasm and willingness to act.

3 **Acceptance** (350) - Acceptance of reality and self.

4. **Reasoning** (400) - Logical thinking and understanding.

Higher levels of consciousness:

1. **Love** (500) - Unconditional love and compassion.

2. **Royalty** (540) - A deep sense of happiness and fulfillment.

3. **Peace** (600) - Inner peace and harmony.

4. **Enlightenment** (700-1000) - A state of supreme consciousness and oneness with the universe.

Each of these levels represents different emotional and energetic states that affect our lives and the way we perceive the world.

Interestingly, according to Hawkins, most of society stops at the Pride level, that is, 175 units on the Hawkins scale, although there is a real opportunity for everyone to develop and can through their development reach much higher levels. People with a very high development of consciousness can even reach the level of Love, that is 500 units on the Hawkins scale. The most outstanding individuals of the Buddha, Jesus' type have reached the level of Enlightenment, that is up to 1000 units. What's interesting is that with development, somewhere above 500 units, a person gets rid of all his unfavorable physical and mental ailments, or diseases. His life is fulfilled, he comes out of the cycle of samsara and achieves nirvana.

A certain analogy to this is the theory of Relative Moral Level Factor presented by Dr. Jan Pajak that humans achieve. The analogy is almost identical. According to Jan Pajak, the quality of life is determined by the aforementioned Relative Moral Level Factor µ, which ranges from 0<µ<1

And so, for

$0<\mu<0.3$ - a person begins to get sick. He begins to have health problems, his status social status rapidly decreases, he simply simply begins to lack Moral Energy.

$0.3<\mu<0.5$ - the human situation is much better, improving in every aspect of human activity.

$0.5<\mu<0.6$ - a person is happy, well-liked, has no enemies. He enjoys excellent health.

$\mu>0.6$ - man falls into euphoria, comes out of the cycle of birth (samsara), achieves nirvana.

The analogy between the two ideas can be seen here, which only confirms the validity of both concepts. Besides, they have been confirmed by reliable esotericisms in the indications of the respective pendulums. It is not without reason that these two ideas are invoked, because we are about to use them in the appropriate settings of the Philosopher's Stone Algorithms, so as to change our Base Vibration and our Relative Vital Energy Level in the desired manner. For what purpose? For the improvement of our health and well-being, for our growth on the Path of Spiritual Development, which is, after all, the meaning of our existence here on the Measurable Level.

Although these scales show a very high analogy, it is worth remembering that they probably can't be applied so one-to-one. Because, for example, a value of $\mu > 0.3$ probably corresponds to a vibration level of 200 units on the Hawkins scale, but already a Love Level of 500 units on the Hawkins scale may already correspond to $\mu > 0.5$. And then, the higher on both scales, the tighter the analogy.

With the Physiological Stone algorithm, you can increase both the Vibration Level and the Relative Moral Energy Level. Why? Because with the Algorithm of the Philosopher's Stone we influence and change the Source Causal and Purposeful Causes behind the given problem, issue, and this is an activity in accordance with the laws of Physics (esoPhysics), and not with some mere wishful thinking. Only, I would rather caution against such throwing oneself immediately into deep waters. And I would rather advise, if one has a Base Vibration of 200, not to set the Algorithm to go from 200 to immediately 500 units on the Hawkins scale, analogously this also applies to µ. Why, because we may not be able to withstand it and it will be beyond the endurance of our "entities". It is better to make changes, but small ones. For example, you can boldly make the adjustment of the Algorithm at once, the transition from 200 to 220 Hawkins units, possibly from µ=0.33 to µ=0.35. And so gradually, in the long run, make positive changes. Without shock therapy, as it were. Here is my proposal for setting the Philosopher's Stone Algorithm to go from a Vibration level of 200 to 220 units on the Hawkins scale, and at the same time go from µ=0.33 to µ=0.35.

-

[Incant (enchant?) in thought or whisper].

...I peel off point No.1, this is the point associated with my right thumb, this is the point associated with (-), this is the actual state that exists.

By the power of My Free Will from My Transcendence....

...I assign to point no.1 a state function that defines and expresses ...

My Base Vibration on the Hawkins scale is 200 units, which corresponds to the Courage Level. And my Relative Moral Energy Level fluctuates around $\mu=0.33$. I feel some discomfort with this, complaining of periodic illness, fatigue and life-weariness.

...

... By the Power of My Free Will from My Transcendence, I place Item No.1 on the

Measurable Level and the Non-Measurable Level....

[This is the first step of the Algorithm].

Another second step of the Algorithm:

[Incant].

...I peel off point No.2, this is the point associated with my left thumb, this is the point associated with (+), this is the state I desire.

By the power of My Free Will from My Transcendence, I assign pointwise No.2 to the state function, which is determined by the Point in the Divine Energy Matrix (Healing), which means....

My Base Vibration is strengthened and increased to 220 units on the Hawkins scale. My Relative Moral Energy Level is increased to $\mu=0.35$. This is accompanied by a significant increase in my internal energy. I feel better, I get sick less often. I am more optimistic about the world, life and the world is more favorable to me.

...

[Incant].

... this Point in this Divine Energy Matrix chosen by me determines the solution (+) to this problem (of Point No.1), determines the state function that I by the Power of My Free Will assign to Point No.2.

By the power of My Free Will, I place Item No.2 on the Measurable Level and the Non-Measurable Level....

[This was the second step of the Algorithm].

The next, third, step of the Algorithm:

[Incant].

... By the Power of My Free Will, I am now performing quantum entanglement of point No.1 with point No.2 due to the state functions of these points.

By the power of My Free Will, I place such quantum entangled points No.1 and No.2 in the Sacred Space of My Heart in the Field of My Heart....

...let There, with the coherence of this quantum entanglement preserved at all times, permanent and positive quantum processes occur for me, consistent with the intention of this entanglement. If there secretes in an additional way negative energy that could harm me or others, by the Power of my Free Will I ground it in a safe way....

(A period of passive waiting follows, which can last quite a long time. However, if there is a sudden breakthrough, perceived as some kind of spasm, chills, vibration or other clear body signals we end this phase and go straight to the last step of the Algorithm).

The final step of the Algorithm:

[Incant].

...By the Power of My Free Will I am now performing a decoherence (breaking) of the quantum entanglement between points No.1 and No.2

If, in the process, I emit additional negative energy that could harm me or others, by the Power of My Free Will I ground it.

(After completing the Algorithm, you shake your hands towards the ground and clap to end the session).

It is not without reason that I have devoted one chapter of this book to radiesthesia and pendulum work, because it is with the help of a pendulum that we can determine our Base Vibration Level and our Relative Moral Energy Level µ. With the help of a pendulum, for example, Izis or another pendulum, we can ask our intuition other necessary questions in the course of setting up the mental Quantum Tools, and we can expect specific guidance through this path, precisely from our intuition.

We will also use this patent in the rest of the book, for further settings of Philosopher's Stone Algorithms (Cognitive Prothesis).

c. Psychological treatments

Well, as you've probably figured out by now this whole book is about quantum psychology in this latest modernist version of esoPhysics. So it's worthwhile now to present some "settings" of Cognitive Proteism relevant to this content.

Here we have a case of a man whose nerves are "eating him alive", unable to cope with the flood of unfavorable and destructive emotions, feelings, thoughts. A man who doesn't know how to get rid of depressing ruminations, that is, over and over again analyzing all his imaginary and unimaginative problems. A man who has not experienced any help from loved ones, friends and even from the health service for various reasons. The reason for such his situation may be various. Maybe he was hurt, injured, someone failed his trust, or maybe it was his illness that led to this, or maybe loss of livelihood? There could be a multitude of reasons. This man's situation in the long run can lead him to serious psychological consequences, even such that he exhausts his limit of mental resilience, and then it will already be a disaster.

Suppose he is your uncle, and you decided to help him by doing some Cognitive Prosthesis sessions for him. Here is the proposed "setup" for such a Cognitive Prosthesis. You do the prosthesis as usual, under similar conditions as always.

-

[Incant (enchant?) in thought or whisper].

...I peel off point No.1, this is the point associated with my right thumb, this is the point associated with (-), this is the actual state that exists.

By the power of My Free Will from My Transcendence....

...I assign to point no.1 a state function that defines and expresses ...

My uncle Adam is mentally exhausted. He doesn't know how to deal with stress. He has adverse and pathological changes in his nervous structures, his brain structures and even his Transcendence. He has a weakened and dysfunctional his mental system, his emotional system, his emotional system and his sense of security system. Too much Cortisol is secreted in his brain and body, as well as Prolactin, too little of the neurotransmitter GABA is secreted. He can't control his nerves, he's constantly flooded with ruminations, constantly analyzing his imaginary and unimaginative problems. His frustration builds up. He is all the way shaking with nerves, his situation worsens day by day. He lacks mental, emotional and emotional endurance and resilience. Also his dopamine system, control and reward, is dysfunctional. His condition has been deteriorating since he retired five years ago. He doesn't know how to find a place for himself. He was widowed ten years back, and his only son emigrated to the Netherlands and has no contact with him. His Base Vibration is apparently less than 150 units on the Hawkins scale. Which causes his condition to deteriorate progressively.

...

... By the Power of My Free Will from My Transcendence, I place Item No.1 on the

Measurable Level and the Non-Measurable Level....

[This is the first step of the Algorithm].

Another second step of the Algorithm:

[Incant].

...I peel off point No.2, this is the point associated with my left thumb, this is the point associated with (+), this is the state I desire.

By the power of My Free Will from My Transcendence, I assign pointwise No.2 to the state function, which is determined by the Point in the Divine Energy Matrix (Healing), which means....

My uncle Adam is already feeling better mentally. He lacks adverse and pathological changes in his behavior, in his behavioral behavior, in his psyche and in his mental behavior. There are no negative pathological changes in his neural and cerebral structures and in his Transcendence. He has a strong, healthy, powerful mental, emotional, sentimental and sense of security system. His brain does not secrete much Cortisol, and if it does, it is secreted in eustress, simultaneously with the secretion of the neurotransmitter GABA in appropriate and optimal doses and proportions. His brain also does not secrete much Prolactin. Instead, his brain secretes Oxytocin, which heals and revitalizes all nerve and neural damage. Adam is now able to keep his nerves in check, no longer flooded with ruminations, no bad thoughts, no bad emotions, no bad feelings. He is in full control of himself. His brain's dopamine system is fine, the reward and control system. He no longer feels frustrated and depressed. He is cheerful, calm, composed and resigned to his fate. He is trying to make contact with his son Christopher. He is mentally and physically healthy, nothing ails him. He maintains normal mobility for his age. He has reopened to contacts with his sister Eliza's family. His interpersonal relations are normal. His Base Vibration has risen to a level of 200 units on the Hawkins scale. His condition is improving day by day. And this is a positive and lasting change in him and the world towards him.

...

[Incant].

... this Point in this Divine Energy Matrix chosen by me determines the solution (+) of this problem (of Point No.1), determines the state function that I by the Power of My Free Will assign to Point No.2.

By the power of My Free Will, I place Item No.2 on the Measurable Level and the Non-Measurable Level....

[This was the second step of the Algorithm].

The next, third, step of the Algorithm:

[Incant].

... By the Power of My Free Will, I am now performing quantum entanglement of point No.1 with point No.2 due to the state functions of these points.

By the power of My Free Will, I place such quantum entangled points No.1 and No.2 in the Sacred Space of My Heart in the Field of My Heart....

...may There, with the coherence of this quantum entanglement preserved at all times, permanent and positive quantum processes occur for my uncle, consistent with the intention of this entanglement. If there secretes in an additional way negative energy there, which could harm me, harm Adam or harm others, by the Power of My Free Will I ground it in a safe way....

(A period of passive waiting follows, which can last quite a long time. However, if there is a sudden breakthrough, perceived as some kind of spasm, chills, vibration or other clear body signals we end this phase and go straight to the last step of the Algorithm).

The final step of the Algorithm:

[Incant].

...By the Power of My Free Will I am now performing a decoherence (breaking) of the quantum entanglement between points No.1 and No.2

If, in doing so, I give off additional negative energy that could harm me, my Uncle Adam harm me or harm others, by the Power of My Free Will I ground it.

(After completing the Algorithm, you shake your hands towards the ground and clap to end the session).

-

This, of course, is just an example of setting the Cognitive Prosthesis. After all, the algorithm Philosopher's Stone or the Cognitive Prosthesis can be set for oneself or for another person with the same effectiveness. Because, as I have already written mental Quantum Tools express "work" on the Unmeasurable Level and are an expression of the action of human Free Will. This appears to be true Magic, but wherever we take seriously human Free Will as an elementary force and Causal Cause from the Unmeasurable Level, we will always be exposed to the charge of practicing Magic. Perhaps this is why the official Science Current is so skeptical of similar revelations that I put in my books. So you can see in this example how the cognitive prosthesis, its setting affects all psychological and behavioral processes of a person, determines the quality of his experiencing emotions and experiencing feelings, experiencing and somehow his thoughts. So, in this example you can see how we with our emotions, our feelings sculpt our soul, our spirit. This is how the path of spiritual development is carried out in him. Its quality is indicated by our well-being, our thoughts, our emotions, our feelings. Distortions in this segment greatly affect the sculpting of this very thing that we do on our psyches, our consciousnesses, our souls. This is, of course, a complicated process, it is difficult to describe it as a single source, it is difficult to characterize this process in simple cannons, this complex process, and it determines not only our quality of life here on earth, the one on the Measurable level. But precisely, as I mentioned, it carves our soul, our spirit, and with this carving we after death get settled. What we carve we are left with final and accounted for later in our lives. On this basis, the Cognitive Protein can be set in a multitude of ways, different or the same processes can be considered for it, but this is obviously not the reason to play. Only behind this there is a deeper reason, precisely deeper for our path of spiritual development, why we came here to earth to this Measurable level, to live this life, to experience the emotions of thoughts and feelings. Someone will say: right away, right away, but it is said that negative emotional experience ennobles a person. To some extent, this is certainly correct, but now for quantum man, homo sapiens quantum entanglement we can closely control the process. We are no longer condemned to such a godly life, to live as their fate will bring it. Nowadays we ourselves can create our own dole in this way, we can face adversity and to a large extent remedy the negative processes of our in our psyches, our emotions our feelings. Before I go into further examples of the findings of the cognitive prosthesis, it is worth pointing out that all of these settings arise from basic quantum laws, arise from the comparison of human Free Will as the causal force and causal cause that shapes our fate, from the acknowledgement that we ourselves can influence whether it affects us. We are not so passively doomed to all that other people will do to us, we can take up arms against it, we can fight something for it, we can overcome it, of course, to a certain extent, because there are certainly some changes that cannot be straightened out. Just as a person who has lost a leg will no longer grow back another leg, certain mental processes on the other hand are irreversible, but on the other hand we can change a lot, more than we think, and this is optimistic, it gives us hope. Now occurs is this phase transition into the era of quantum man. Such processes will be normal acceptable. Now they may seem overly fanciful, they may seem too related to magic, but even if someone accuses me of such magical practices, however, this magic of mine is white magic, it is not Magic that harms anyone, it is Magic that is meant to help the people concerned, those who are affected precisely by these internal, mental, emotional, emotional issues. I'm about to try to show the setting of a cognitive prosthesis for people living under unimaginable stress, living at the mercy of the media, people doomed to Elvis Presley syndrome, people who suffer great mental anguish because of too much interest and self-interest in their lives turns out that even them can strongly help, actually before very deep behavioral changes, big psychological troubles because people who are subject to this syndrome of too much interest in the media and other people's lives, so far have had no real preventive measures for this, given means to free themselves from such a situation of permanent stress. So, imagine that you are such a popular person that the media torments and torments you, and the prying of interested people about your life is unbearable.

Fame and popularity

I now turn to the delicate matter of fame and popularity. This, too, depends on a person's predisposition and some bear it better others worse. However, psychology is quite unequivocal about the impact of fame on a person's mental state and condition. It is not recommended, which sounds strange to those involved, that is, those bigger or smaller celebrities and people who necessarily and at all costs want fame. Why? Because, although admittedly there is that element of the Spiritual, of Transcendence in each of us, our brain, like the whole body, is a product of evolution and has its limitations from this point of view. Humans evolved from small groups of communities, a tribe, a tribe, a tribe, which had a maximum of 300. tribesmen. And this is the number of acquaintances almost the maximum that the average person can have. Anything above that is downright harmful to our psyches and negatively affects our spiritual development. Above this magic number of 300th acquaintances the torment begins, the suffering begins. Still, if someone has such fame that it brings him almost only profits and people, his fans, adore him without a shadow of envy, well, you can still somehow live. But!!! But when something begins to break down in our lives, and "our" fans begin to be aware of it. Some divorces, quarrels between celebrity partners, illnesses, some emotional crises. And when it turns out that we are ordinary people with our advantages and disadvantages, the reaction of "fans" in a face-to-face relationship, when they meet their idols, can be really cruel, because, after all, idols, these "gods" of our time, must be perfect and must lead perfect lives. And these are from their (from the fans' side) some unconscious, uncontrolled, even one could say: subliminal behavior. Ultimately, however, they give such a push to some famous people that they "shy away from everything." There are a few things that can actually glaze the brains of those affected. And one of those things is fame and popularity. That unfavorable fame and popularity. Famous people save themselves from the ill-effects of fame, so understood, by the fact that they live every day in certain enclaves to which only a select few have access. They don't so much isolate themselves from the public, but rather limit others' access to themselves. There is an opinion that public figures, however, face great inconveniences in their lives. Practically nothing then defends their privacy. Very telling then is the attitude of media representatives, who believe that famous people and people in public positions are virtually deprived of their rights to privacy. The media practically "persecute" such people. Without any restraint they publish materials from their lives, even very intimate ones. And this, to a large extent, very negatively affects the mental condition of such, what not to say, "persecuted" people. This is not discussed; these people are not "defended". Others, the "ordinary" ones, think that since the celebrities have money and other profits, let them not protest and pity, because thousands of other people would like to be in their place. However, people, these "ordinary" people, these ordinary bread eaters, do not realize how aggravating and "painful" such a life is in the long run for those on the front pages. Everything is blamed on these limitations of our physical brains, our psyches. These limitations due to the physicality of our beings. Some of these public figures end up very badly, of which there are more than enough examples. However, I won't mention anyone by name here, because I don't want to expose myself to criminal liability and violation of the personal rights of such people. I can only refer here to the fact that as a writer I use the poetics license here and am protected. Those who want to, believe it, those who don't, don't believe the content I write here. I do not force anyone; I do not claim that I am always right.

But I advise everyone, however, to think deeply before such a choice of "public" life.

But fortunately, in the age of Homo Sapiens Quantum, there are already appropriate methods, or Quantum Tools, that allow you to "cope" with these inconveniences of the life of a public figure, a celebrity, or that allow you to help yourself a lot in this regard at the time. What should be done in such a case? In the event that we have to "endure" such a life? We must properly "set" and consequently use such a Quantum Tool. In this case, let's call such a tool the Cognitive Prosthesis.

[Incant (enchant?) in thought or whisper].

...I peel off point No.1, this is the point associated with my right thumb, this is the point associated with (-), this is the actual state that exists.

By the power of My Free Will from My Transcendence....

...I assign to point no.1 a state function that defines and expresses ...

I am a popular person (actor?). I have fans all over the world who have been actively following my life for many years. For the past few years, however, my situation has become unbearable for me. I have become a constant source of national and foreign media attention. It even haunts me. This is accompanied by concrete negative pathological changes in my brain and nervous system. This is accompanied by concrete negative behavioral changes in my behavior, even in my character and in my psyche. This is accompanied by concrete negative changes and destruction in my emotional system, emotional system, logical system, security system in my brain. This is accompanied by concrete negative changes and destruction in my behavior, in my behavioral behavior, in my psyche, in my mental behavior, in my nervous structures, in my brain structures in my Transcendence (Higher Self). Sometimes I already think I'm on the verge of a mental breakdown. My system in my brain: Prefrontal lobes- Frontal lobes- Amygdala - Hippocampus is overloaded, overly negatively stimulated. Too much cortisol is secreted in my brain, not enough of the neurotransmitter GABA is secreted. I feel that from stress and negative emotions I'm even getting carried away. There is no longer a positive correlation and interdependence in the cortisol and GABA neurotransmitter secreted in my brain and body. I don't sleep through the night. The worst part is that I have to shine my eyes in front of people, my family and my friends for what the media is doing to me, what I myself am doing because of the negative behavioral changes in my behavior. My situation is getting worse year after year, month after month.

...

... By the Power of My Free Will from My Transcendence, I place Item No.1 on the Measurable Level and the Non-Measurable Level....

[This is the first step of the Algorithm].

Another second step of the Algorithm:

[Incant].

...I peel off point No.2, this is the point associated with my left thumb, this is the point associated with (+), this is the state I desire.

By the power of My Free Will from My Transcendence, I assign pointwise No.2 to the state function, which is determined by the Point in the Divine Energy Matrix (Healing), which means....

I already feel much better mentally. I already have my nerve structures calmed down, my brain structures, my cerebellum, my sympathetic trunk, I already have my whole mind calmed down. I feel mental balance, I am far from having a mental breakdown. I have a strong, healthy, powerful emotional system in my brain, I have a strong, healthy, powerful emotional system and logical system, I have a strong healthy, powerful, healthy sense of security system. My system in my brain: Prefrontal Lobes- Frontal Lobes- Amygdala-Body-Hippocampus is working well, without any dysfunction or abnormality. In my brain, the dopamine system, the reward system and the control system are working well and harmoniously. Dopamine and serotonin are secreted in a normal, healthy and harmonic way. I have a calmed nervous system, nerves. Not much cortisol is secreted. Cortisol is only secreted in an optimal and healthy way in proper correlation with the secreted neurotransmitter GABA. I am completely unconcerned anymore about what the press writes about me, what they say about me in the media. I have infinite mental resilience, I have infinite emotional resilience, I have infinite emotional resilience, I have infinite logical resilience, and I have infinite resilience, in terms of my sense of security, to what the media, my enemies, people who are not favorable to me, are doing or will do to me and my family. I have infinite mental strength in myself, I have infinite emotional strength in myself, I have infinite emotional and logical strength in myself, I have infinite strength when it comes to my sense of security in myself against what is happening negatively or will happen negatively to me and my family now and in the future. I am mentally and physically healthy. I feel good mentally and physically. My Base Vibration on the Hawkins scale is at 300 units on this scale. I don't have a muddled mind; I don't have an overloaded mind. I feel light, spring-like. I am satisfied with life. I'm enjoying life. And these are positive and lasting changes in me and the world toward me.

...

[Incant].

... this Point in this Divine Energy Matrix chosen by me determines the solution (+) to this problem (of Point No.1), determines the state function that I by the Power of My Free Will assign to Point No.2.

By the power of My Free Will, I place Item No.2 on the Measurable Level and the Non-Measurable Level....

[This was the second step of the Algorithm].

The next, third, step of the Algorithm:

[Incant].

... By the Power of My Free Will, I am now performing quantum entanglement of point No.1 with point No.2 due to the state functions of these points.

By the power of My Free Will, I place such quantum entangled points No.1 and No.2 in the Sacred Space of My Heart in the Field of My Heart....

...let There, with the coherence of this quantum entanglement preserved at all times, permanent and positive quantum processes occur for me, consistent with the intention of this entanglement. If there secretes in an additional way negative energy that could harm me or others, by the Power of my Free Will, I ground it in a safe way....

(A period of passive waiting follows, which can last quite a long time. However, if there is a sudden breakthrough, perceived as some kind of spasm, chills, vibration or other clear body signals we end this phase and go straight to the last step of the Algorithm).

The final step of the Algorithm:

[Incant].

...By the Power of My Free Will I am now performing a decoherence (breaking) of the quantum entanglement between points No.1 and No.2

If, in the process, I emit additional negative energy that could harm me or others, by the Power of My Free Will, I ground it.

(After completing the Algorithm, you shake your hands towards the ground and clap to end the session).

Of course, this is only a proposal for the "setting" of the Cognitive Protein for this case. Anyone interested is free to compose such an alignment in relation to himself, yes or slightly different. Now it is still necessary in a quiet place, with relaxing music or in silence to maintain this quantum entanglement so about 1.5 hours. That is, the time depends largely on one's talent. Some can perform such a séance longer others shorter to get the right effect. At the first séance, negative, unwanted and painful emotions, feelings, irritations may flow out.

This should rather be endured, these bad energies simply release and flow away from us, from our brain, mind, they will no longer harm us. As I wrote the real gains come only after a series of such seances, and this series works according to the progression of ½;¼;1/8;1/16;1/32; etc. But you can see that after just four séances performed well, that is, performed optimally, we should get rid of the problem to 1/16 of the initial value, and this is already a lot.

It is also worth noting that it is necessary to maintain a certain interval between séances so that the quantum processes taking place can be fully realized. I use at least a day of such an interval between séances, but, of course, this is an empirical matter, so each for himself must determine such an interval time between séances.

Whether we want it or not, we have this legacy of animalism in our brain, in our mind, which strongly projects on us, on our Soul (Transcendence). We will not free ourselves from this, it is worth knowing at a certain age our limitations in this regard. This animalism is necessarily necessary for us in this "life" on this Measurable Level, in which our entire physicality, our senses, function.

Another social group that, lo and behold, will now be able to resist a kind of psychological violence, will be all those persecuted because of their skin color, social background, or even caste in societies where such problems still exist. They will be able to resist psychological violence those from poorer homes, those less educated, those outside the system. This is not talked about out loud, but all sorts of barriers still exist and are doing well even in European societies, not to mention Asia or poor Africa. Sometimes they effectively clip the wings of ambitious people who are just that little bit unlucky to be outside the arrangements.

On these patterns that I have shown in this book, anyone with a little good will can "set up" such a Cognitive Prosthesis for himself. And now I propose to those interested themselves to perform such an exercise. Let me remind you: we set up a problem, then the solution to that problem, and we quantum entangle the whole thing in the Heart Field.

With this last example, I would like to conclude this brief discussion of the Cognitive Prosthesis in this its psychological guise. After all, the Cognitive Prosthesis, and the Philosopher's Stone Algorithm in general, can still be applied to specific pains and diseases of the body and soul. I have discussed this in part in my earlier publications. I wish everyone to find the optimal setting for the diseases that affect him personally.

7 Threats

EsoPhysics, or rather, the practical consequences arising from it, shows that indeed Magic is the result of the application of quantum laws. In other words, it can even be said that this Traditional Magic is justified. That being the case, the question is, why has Magic always been fought against by the official mainstream of science, rather than lavished and developed? The answer is because it is and has been dangerous, and one has never been able to "control" it in such a way as to control it. Even before the existence of the Age of Enlightenment, which flagrantly and transparently fought darkness, quackery and superstition (read Magic), all manifestations of magic were already fought. For the Church in particular, it was a prime opponent. But with the advent of more enlightened centuries, Magic became all the more the proverbial whipping boy. From those days until today, the situation has become entrenched, and even exacerbated, as a result of the fact that the prevailing scientific Interpretation (strictly: interpretations) of Quantum are strictly materialistic, atheistic and single-level, and recognize only strict Physicalism. That is, what the measuring devices do not indicate, it does not ontologically exist. Unfortunately, recent scientific currents have begun to admit cautiously that perhaps there is something else that does not yield to Physicalism. The issue is precisely about quantum entanglement and the observation that in quantum entanglement quantum leaps cannot, after all, take place against Albert Einstein's General Theory of Relativity under the same conditions as all other physical processes. My answer and proposal to solve these paradoxes is esoPhysics and its fundamental division of reality into two levels. Into a Measurable Level and an Unmeasurable Level. I will not here, of course, summarize all of esoPhysics with its solutions to the major paradoxes of old Quantum Mechanics. However, it is worth realizing that esoPhysics sanctions and confirms the existence and actions of Magic, which I call Scientific Magic. Moreover, esoPhysics shows on what principle today the workings of Magic can be explained. By the same token, however, it gives a weapon to anyone familiar with esoPhysics to fight against all the negative consequences of such Magic. It is possible, and this is a strongly optimistic approach, to successfully oppose negative Magic, say it: Black Magic. So, to speak, by applying an antidote of action to this bad Magic. This is so within the framework of what are the dangers of applying quantum laws.

Well, there they are, it is this Evil Scientific Magic. But right away it is an antidote. It will be the same Scientific Magic, but used for a good purpose, by us. People get at their disposal not only mental Quantum Tools, but also a weapon against Evil Traditional Magic. Because it is not talked about and talked about loudly in the circles of the mainstream Science, but magic has always done well and flourished, despite the fact that it was not taken seriously by science. And particularly dangerous have always been the

Black Mages and witches, who carried out their practices with complete almost impunity, in the sense of accountability to the law. This was probably also the one main reason that Science was so separated from everything that could not and could not be measured. But it's time to stop burying our heads in the sand. It is time to face the full truth in the age of Homo Sapiens Quantum. It is necessary to take for granted the postulates of esoPhysics. Because what 20th century Science is trying to serve us is true darkness and superstition, with the addition of scientific superstition. It is the aftermath of paganism stuck in us all the time. ...The truth will set you free.... -in the words of the Bible.

Yes, as I wrote, wherever the action of Free Will of man as a new Elemental Force and Source Causal Cause is allowed, there we are dealing, whether we want it or not, with Magic. At the same time, it will be Traditional Magic, when we treat quantum laws intuitively, or it will be Scientific Magic, Quantum Magic, when we deliberately use quantum laws including quantum entanglement, that most magical of all quantum laws of Nature.

And while Science has never recognized Magic as some kind of key element of the scientific narrative, this has not bothered the Magi and Sorceresses in any way. If it were only Magic limited to White Magic, there would be no problem. However, Mages accustomed to their impunity cross the line and litters evil Magic against innocent people. People have so far been deprived of any protection against such evil. Therefore, all that was left for them was passive obstruction and hope that they would not fall under such action. Today, however, now that there is a method, it is worthwhile to set oneself such a Philosopher's Stone Algorithm from time to time to guard against such unexpected and hidden aggression. Here I will now propose to set up a mental Quantum Tool (Philosopher's Stone Algorithm) against the hidden attack of black magic on us.

: Sit somewhere in a comfortable posture or lie on your back. Try to force your family not to disturb you during this time (a good few tens of minutes). You can be accompanied by quiet relaxing music (Chillout music) in the background.

[Incant (enchant?) in thought or whisper].

...I peel off point No.1, this is the point associated with my right thumb, this is the point associated with (-), this is the actual state that exists.

By the power of My Free Will from My Transcendence....

...I assign to point no.1 a state function that defines and expresses ...

I intuitively read that I have been a victim of black magic against me or my immediate family members for some time. Because of this, I have my mental system weakened, I have my emotional system weakened, I have my emotional system weakened, I have my logical system weakened. I have some kind of strange bad streak in my personal life, but also in my professional life. Despite my efforts, all my efforts seem to be misguided and ineffective, clearly someone is wishing me badly. Some kind of curse is weighing down on my fate, someone is deliberately messing with my fate in this negative sense. And I am completely helpless and observe in myself also increasing health problems. This also has a very negative impact on my well-being.

...

... By the Power of My Free Will from My Transcendence, I place Item No.1 on the Measurable Level and the Non-Measurable Level....

[This is the first step of the Algorithm].

Another second step of the Algorithm:

[Incant].

...I peel off point No.2, this is the point associated with my left thumb, this is the point associated with (+), this is the state I desire.

By the power of My Free Will from My Transcendence, I assign pointwise No.2 to the state function, which is determined by the

Point in the Divine Energy Matrix (Healing), which means....

I have completely neutralized the effect of black magic against me and my family. I submit myself to the full protection of Archangel Michael in this regard. I have a strong, healthy, functioning of my system: mental, emotional, emotional, logical. Anyone who litters magic against me or my family does not harm me, does not hurt me, does not do me any evil. Such a person does not harm me or my family, he only harms himself, and in a powerful way. All this doesn't negatively affect my psyche, it doesn't negatively affect my emotions, it doesn't negatively affect my feelings, and it doesn't negatively affect my karma. It all flows down to me, like water flows down over a duck, it is completely indifferent to me. I am feeling very well. I am healthy mentally and physically, and the mages stay away from me and my family. And this is a positive and lasting change in me, and the world towards me.

...

[Incant].

... this Point in this Divine Energy Matrix chosen by me determines the solution (+) of this problem (of Point No.1), determines the state function that I by the Power of My Free Will assign to Point No.2.

By the power of My Free Will, I place Item No.2 on the Measurable Level and the Non-Measurable Level....

[This was the second step of the Algorithm].

The next, third, step of the Algorithm:

[Incant].

... By the Power of My Free Will, I am now performing quantum entanglement of point No.1 with point No.2 due to the state functions of these points.

By the power of My Free Will, I place such quantum entangled points No.1 and No.2 in the Sacred Space of My Heart in the Field of My Heart....

...let There, with the coherence of this quantum entanglement preserved at all times, permanent and positive quantum processes occur for me, consistent with the intention of this entanglement. If there secretes in an additional way negative energy that could harm me or others, by the Power of my Free Will, I ground it in a safe way....

(A period of passive waiting follows, which can last quite a long time. However, if there is a sudden breakthrough, perceived as some kind of spasm, chills, vibration or other clear body signals we end this phase and go straight to the last step of the Algorithm).

The final step of the Algorithm:

[Incant].

...By the Power of My Free Will I am now performing a decoherence (breaking) of the quantum entanglement between points No.1 and No.2

If, in the process, I emit additional negative energy that could harm me or others, by the Power of My Free Will, I ground it.

(After completing the Algorithm, you shake your hands towards the ground and clap to end the session).

This is just such an example, and probably many of you would "set" such a mental Quantum Tool a little differently. So, I only encourage this attitude of constructing your own "settings" of the Algorithms Philosopher's Stone. There are many channels on YouTube.com that offer specific music and video files dedicated to this very issue of fighting black magic. They work on the principle of Resonance, so the power of their action is definitely greater.

Table of Contents:

Introduction

1.Quantum Psychology. Kickoff

2. Traditional take

3. Stress

4. Strange Energy

5. Menatal quantum tools

6. Practical application of mental Quantum Tools

a. Psychological prevention

b. Maps of consciousness, Relative Moral Energy Level µ.

c. Psychological treatments

7. Threats

www.ingramcontent.com/pod-product-compliance
Lightning Source LLC
Chambersburg PA
CBHW052149220526
45471CB00004B/1588